Lecture Notes in Mathematics

Edited by A. Dold and B. Eckmann

742

Kevin Clancey

Seminormal Operators

Springer-Verlag
Berlin Heidelberg New York 1979

Author

Kevin Clancey
Department of Mathematics
University of Georgia
Athens, GA 30602
USA

AMS Subject Classifications (1970): 47B20, 47A65

ISBN 3-540-09547-0 Springer-Verlag Berlin Heidelberg New York
ISBN 0-387-09547-0 Springer-Verlag New York Heidelberg Berlin

Library of Congress Cataloging in Publication Data
Clancey, Kevin, 1944-
Seminormal operators.
(Lecture notes in mathematics; 742)
Bibliography: p.
Includes index.
1. Subnormal operators. I. Title. II. Series: Lecture notes in mathematics (Berlin); 742.
QA3.L28 no. 742 [QA329.2] 510'.8s [515'.72] 79-20324
ISBN 0-387-09547-0

Printing and binding: Beltz Offsetdruck, Hemsbach/Bergstr.
2141/3140-543210

Preface

These notes are concerned with seminormal operators on a Hilbert space. In the past decade several major results on this class of operators have been obtained, some of which appear mystifying and have created stirs of interest in the area. These results have come from (at least) five different sources and this makes it somewhat difficult to appreciate what is happening. The aim of these notes is to paint a reasonably self-contained picture of some of the developments in the area of seminormal operators which have occurred during the last ten years.

The fact that seminormal operators are interesting became obvious in 1970 when C. R. PUTNAM [5] established that the planar Lebesgue measure of the spectrum of any non-normal seminormal operator is positive. Perhaps just as deep is the result obtained in 1973 by C. A. BERGER and B. I. SHAW [1] which shows that any hyponormal operator with a cyclic vector has a trace class self-commutator. While these two developments were taking place R. W. CAREY and J. D. PINCUS were studying an invariant (referred to as the principal function) for operators having a trace class self-commutator. This invariant arose as a two dimensional analogue of the phase shift from the theory of trace class perturbations of self-adjoint operators. HELTON and HOWE [1], in an attempt to understand the work of Carey and Pincus, began studying the star-algebra generated by polynomials in an operator that has a trace class self-commutator. These authors introduced a tracial bilinear form on this algebra which they represented via integration up against a signed measure on the plane. Pincus immediately verified this measure has derivative equal to the principal function. Independently, these authors have used these invariants to describe a fairly complete spectral theory for seminormal operators. A most interesting result concerning these matters is the recent work of C. A. BERGER [1] which shows that the size of the principal function for a hyponormal operator can give information concerning cyclic vectors. That such information was carried in the principal function was conjectured by Helton and Howe. In another direction, J. G. STAMPFLI [8] has produced an interesting dichotomy in the local spectral theory of seminormal operators. Stampfli has shown that local spectral subspaces of hyponormal operators are always closed (but possibly trivial), whereas in the cohyponormal case the local spectral spaces are always non-trivial. It is the above mentioned results of these authors that occupies the major portion of these notes.

The notes are organized as follows.

Chapter 1 is mainly concerned with the local spectral theory of seminormal operators. Examples and simple applications of local spectral theory are presented.

Chapter 2 contains a "singular integral" model for seminormal operators. This model plays an important role in the remaining portion of the notes. In this chapter the model is used to make transparent a pair of self-adjoint commutator inequalities of Putnam [2] and Kato[2].

Chapter 3 contains what its title describes. First, we derive Putnam's inequality which establishes that the planar Lebesgue measure of the spectrum of a non-normal seminormal operator is positive. Secondly, we derive the result of Berger and Shaw [1] which establishes that a hyponormal operator with a cyclic vector has a trace class self-commutator.

Chapter 4 presents a discussion of the phase shift of M.G. Krein [1]. This phase shift arises in connection with trace class perturbations of self-adjoint operators. The existence and properties of the phase shift are crucial to our proof of the existence of the principal function. Several remarks concerning the phase shift are presented which are intended to give the reader a better feeling for the principal function.

Chapter 5 contains a brief study of nearly normal operators. For a portion of this chapter we restrict to the seminormal case. This provides several simplifying advantages. One such advantage is the ease with which we can compute the principal function for singular integral representations of seminormal operators. The finale of the notes is a result of Berger [1] relating the size of the principal function to the existence of cyclic vectors.

A general "thank you" is offered to my colleagues, students, and friends who have influenced the writing of these notes. More specific thanks are given to Ann Ware and Dianne Byrd for their careful typing of this work, and to Tom Howe for a final proofreading. Finally, a special thanks to my wife Carolyn for her constant support.

K.C.

Athens, Georgia
Spring 1979 .

CONTENTS

CHAPTER I

SEMINORMAL OPERATORS

In this chapter, we set down some of the basic properties of seminormal operators. A fairly complete description of the local spectral theory for seminormal operators is presented. Examples are constructed and spectral theory is illuminated with these examples.

1. Definitions and Basic Properties.

The notation $\mathcal{H}$ is reserved for a complex separable Hilbert space with inner product (,). The term operator indicates a bounded linear operator on $\mathcal{H}$. Operators will be denoted by capital letters A, B, C, ... The full algebra of operators on $\mathcal{H}$ is denoted by $L(\mathcal{H})$. As is customary, if A, B are operators on $\mathcal{H}$, then their commutator is denoted by $[A,B] = AB - BA$. The commutator $[A^*,A] = A^*A - AA^*$ is referred to as the self-commutator of the operator A.

As the reader surely knows, an operator $N \in L(\mathcal{H})$ is called normal when N commutes with N^*. Equivalently, N is normal when the self-commutator $[N^*,N]$ is zero. A generalization of normality is the following: An operator $S \in L(\mathcal{H})$ is called seminormal in case its self-commutator $D = [S^*,S]$ is semidefinite. In the case $D \geq 0$ the operator S is called hyponormal and when $D \leq 0$ the operator S is called cohyponormal. In spite of the fact that the adjoint of the class of hyponormal operators is merely the class of cohyponormal operators, there are several important stages where the development of the classes is entirely different. Examples of these differences appear in local spectral theory and in the study of cyclic vectors.

For any $A \in L(\mathcal{H})$ and $\alpha, \beta \in \mathbb{C}$

$$[(\alpha A+\beta)^*, \alpha A+\beta] = |\alpha|^2 [A^*, A] .$$

Consequently, $\alpha S + \beta$ will be seminormal whenever S is seminormal. This last remark remains valid if in both instances the word "seminormal" is replaced by either "hyponormal" or "cohyponormal".

Let $A \in L(\mathcal{H})$ and $z \in \mathbb{C}$. The notation $A_z = A - zI$ will be used. To avoid confusion, we indicate that $A_z^* = (A_z)^* = (A^*)_{\bar{z}}$. Let $A = X + iY$ be the <u>Cartesian form</u> of the operator A, so that $X = \frac{1}{2}[A+A^*]$ and $Y = \frac{1}{2i}[A - A^*]$. Write $z = x + iy$ in Cartesian form. Easy computations show

$$A_z^* A_z = X_x^2 + Y_y^2 + i[XY-YX] \tag{1.1}$$

$$A_z A_z^* = X_x^2 + Y_y^2 - i[XY-YX] \tag{1.2}$$

$$[A_z^*, A_z] = 2i[XY-YX]. \tag{1.3}$$

The identities (1.1)-(1.3) will be used freely in the sequel. Identity (1.3) provides an obvious connection between seminormal operators and pairs of self-adjoint operators with semidefinite commutators.

The following simple lemma will be used.

<u>Lemma 1.1</u>. Let A and B be in $L(\mathcal{H})$. In order that the inequality

$$\|Af\| \leq \|Bf\|$$

hold for every $f \in \mathcal{H}$, it is necessary and sufficient that $A = KB$ for some contraction operator K.

Proof. The sufficiency of the condition is clear. If the inequality $\|Af\| \leq \|Bf\|$ holds, then we define $Kg = Af$, whenever $g = Bf$. This defines K on the range of the operator B, moreover, $\|Kg\| \leq \|g\|$, for g in this range. Thus K extends to a contraction on the closure of the range of B. Setting K to be zero on the orthogonal complement of the range of B provides a contraction operator on $\mathcal{H}$ such that A = KB. The lemma is proved.

The following proposition describes three equivalent formulations of hyponormality:

Proposition 1.1. Let $H \in L(\mathcal{H})$. The following statements are equivalent:

(i) $H^*H - HH^* \geq 0$

(ii) $\|H^*f\| \leq \|Hf\|$, for all $f \in \mathcal{H}$.

(iii) $H^* = KH$, for some contraction operator K.

Proof. Statements (i) and (ii) are obviously equivalent. The equivalence of (ii) and (iii) follows immediately from Lemma 1.1. The proposition is proved.

The following properties of hyponormal operators follow quickly from Proposition 1.1.

1°. Let H be hyponormal. Then we have the inclusion

$$\ker H \subset \ker H^*, \tag{1.4}$$

where for $A \in L(\mathcal{H})$, we have used the notation ker A to indicate the kernel of the operator A. Note, in particular, the inclusion (1.4) shows that ker H is a reducing subspace for the hyponormal operator H.

2°. Let H be hyponormal. Then we have the inclusion

$$R(H) \subset R(H^*), \tag{1.5}$$

where for $A \in L(\mathcal{H})$, we have used the notation R(A) for the range of the operator A. The inclusion (1.5) follows by taking adjoints in statement (iii) of Proposition 1.1 .

From 2° we learn the following. If the equation

$$(H-\lambda)f = \phi$$

has a solution $f = \phi_\lambda$, then the equation

$$(H^*-\bar{\lambda})f = \phi$$

has a solution $f = \Psi_{\bar{\lambda}} = [K(\lambda)]^*\phi_\lambda$ satisfying

$$\|\Psi_{\bar{\lambda}}\| \leq \|\phi_\lambda\| ,$$

where $K(\lambda)$ denotes some contraction satisfying $H_\lambda^* = K(\lambda)H_\lambda$. We note that $K(\lambda)$ is not uniquely determined (this can be remedied by insisting that $K(\lambda)$ be zero on $[R(H_\lambda)]^\perp$). Further, the dependence of $K(\lambda)$ on the parameter λ is not simple.

3°. Let H be an invertible hyponormal operator. The operator H^{-1} is also hyponormal. In fact, if $H^* = KH$, then $K = H^*H^{-1}$. Thus $K^* = (H^{-1})^*H$, which provides the equation $(H^{-1})^* = K^*H^{-1}$. The hyponormality of H^{-1} follows from Proposition 1.1. For another proof of this remark, we could compute

$$[(H^{-1})^*,H^{-1}] = H^{-1}(H^{-1})^*[H^*,H](H^{-1})^*H^{-1} .$$

The latter identity can be used to show H^{-1} is hyponormal. This computation also shows that the ranks of the self-commutators of H and H^{-1} are equal. Similar remarks can be made for seminormal and cohyponormal operators.

The following lemma will be used twice in the sequel. Its simple proof is left as an exercise:

<u>Lemma 1.2</u>. Let $\{a_n\}_{n=1}^\infty$ be a sequence of positive numbers which satisfy the relations

$$a_1^2 \leq a_2 \quad \text{and} \quad a_n^2 \leq a_{n-1}a_{n+1}, \; n = 2,3,\ldots \quad .$$

Then

$$a_1^n \le a_n, \; n = 1,2, \ldots \quad .$$

The notation $\sigma(T)$ will be used for the spectrum of an operator $T \in L(\mathcal{H})$. The spectral radius of T will be denoted by $r_{sp}(T)$. Thus

$$r_{sp}(T) = \max\{|\lambda| : \lambda \in \sigma(T)\}.$$

The following proposition gives an example of one of the properties which seminormal operators share with normal operators:

Proposition 1.2. Let S be a seminormal operator in $L(\mathcal{H})$. Then

$$\|S\| = r_{sp}(S) \quad .$$

Proof. We will employ the identity

$$r_{sp}(S) = \overline{\lim_n} \|S^n\|^{1/n} \quad .$$

Without loss of generality it can be assumed that S = H is hyponormal. Let K be a contraction operator such that $H^* = KH$. Set $a_n = \|H^n\|$, $n = 1,2,\ldots$. Then

$$a_1^2 = \|H\|^2 = \|H^*H\| = \|KH^2\| \le \|H^2\| = a_2 \; .$$

Similarly,

$$a_n^2 = \|H^n\|^2 = \|(H^*)^n H^n\| = \|(H^*)^{n-1} K H^{n+1}\| \le \|(H^*)^{n-1}\| \; \|H^{n+1}\| = a_{n-1} a_{n+1} \; ,$$

for $n = 2,3,\ldots$. From Lemma 1.2, we obtain

$$a_1^n = \|H\|^n \le \|H^n\| = a_n, \; n = 1,2,\ldots \quad .$$

Therefore, $\|H^n\| = \|H\|^n$, $n = 1,2,\ldots$. It follows that $r_{sp}(H) = \|H\|$. The proposition is proved.

The following corollary of Proposition 1.2 gives a growth condition on the resolvent of a seminormal operator:

Corollary 1.1. Let S be a seminormal operator. Assume λ_0 is a complex number such that $\lambda_0 \notin \sigma(S)$. Then

$$\|(S-\lambda_0)^{-1}\| = \frac{1}{\operatorname{dist}(\lambda_0,\sigma(S))} ,$$

where dist $(\lambda_0,\sigma(S))$ denotes the distance from λ_0 to $\sigma(S)$.

Proof. From Proposition 1.2

$$\|(S-\lambda_0)^{-1}\| = \max\{|\lambda| : \lambda \in \sigma((S-\lambda_0)^{-1})\} .$$

This latter quantity equals $[\min\{|\lambda-\lambda_0| : \lambda \in \sigma(S)\}]^{-1}$. This ends the proof.

In the remainder of this section we discuss the usual manner for splitting off a maximal normal part from a seminormal operator. The part remaining used to be referred to as "completely non-normal" or "abnormal" part. More recently, the adjective "pure" has been used to describe this part. We will use this newer terminology.

The seminormal operator $S \in L(\mathcal{H})$ is called pure in case, the only subspace reducing S on which S is a normal operator is the zero subspace.

If H is a hyponormal operator on $\mathcal{H}$ and $\mathcal{M} \subset \mathcal{H}$ is an invariant subspace of H, then the restriction $H|_{\mathcal{M}}$ of H to $\mathcal{M}$ is also hyponormal. In fact, assume $H\mathcal{M} \subset \mathcal{M}$ and let P be the orthogonal projection of $\mathcal{H}$ onto $\mathcal{M}$. If K is a contraction operator satisfying $H^* = KH$, then $(H|_{\mathcal{M}})^* = PH^*|_{\mathcal{M}} = PKPH|_{\mathcal{M}}$. The operator PKP (considered here as acting on $\mathcal{M}$) is a contraction and the remark follows. The same remark cannot be made for cohyponormal operators. For example, the restriction of a unitary operator to an invariant subspace can be an isometric non-normal(hence non-cohyponormal) operator. Nevertheless, as the reader may trivially check, the restriction of a seminormal operator to a reducing subspace is seminormal. Finally, we note the following useful remark.

4°. Let H be a hyponormal operator on $\mathcal{H}$. Assume $\mathfrak{M} \subset \mathcal{H}$ is an invariant subspace for H such that the restriction operator $H|_{\mathfrak{M}}$ is normal. Then $\mathfrak{M}$ reduces the operator H.

The analogue of 4° formulated for cohyponormal operators is not true. The verification of 4° is left as an exercise.

The following theorem often enables the study of seminormal operators to be reduced to the case of a pure seminormal operator.

<u>Theorem 1.1</u>. Let S be seminormal operator on $\mathcal{H}$. Denote by $\mathfrak{M}_0(S)$ the smallest subspace of $\mathcal{H}$ reducing the operator S containing the range of $D = [S^*,S]$. Set $\mathfrak{M}_1(S) = \mathfrak{M}_0(S)^{\perp}$. Relative to the decomposition $\mathcal{H} = \mathfrak{M}_0(S) \oplus \mathfrak{M}_1(S)$, the operator $S = S_0 \oplus S_1$, where S_0 is a pure seminormal operator and S_1 is normal.

Proof. Without loss of generality, it can be assumed that S=H is hyponormal. In this case $D = [H^*,H]$ is non-negative semidefinite. It suffices to show that any reducing subspace for H on which the restriction of H is normal must be orthogonal to $\mathfrak{M}_0(H)$. Let $\mathfrak{M}$ be such a subspace and let $f \in \mathfrak{M}$. Then

$$0 = \|Hf\|^2 - \|H^*f\|^2 = (Df,f) = \|D^{1/2}f\|^2 .$$

Thus $Df = 0$, or in other words, f is orthogonal to R(D). This ends the proof.

2. Examples.

In this section we assemble some common examples of seminormal operators.

Note first, on a finite dimensional Hilbert space every seminormal operator S is normal. Indeed, $\text{trace}[S^*,S] = 0$. Since all the eigenvalues of $[S^*,S]$ have the same sign, this forces $[S^*,S] = 0$.

Let N be a normal operator on a Hilbert space $\mathcal{K}$. Assume $\mathcal{H}$ is an invariant subspace of N. The restriction $A = N|\mathcal{H}$ is called a <u>subnormal</u> operator on $\mathcal{H}$. Let P denote the orthogonal projection of $\mathcal{K}$ onto $\mathcal{H}$ and $P' = I-P$ be the complementary projection, then

$$[A^*,A] = (PN^*N-NPN^*)|_{\mathcal{H}} = PNP'N^*|_{\mathcal{H}} .$$

The last operator is clearly positive and this establishes the well known result that every subnormal operator is hyponormal. It is interesting to note that subnormal operators used to be called hyponormal operators. It is not trivial to construct a hyponormal operator which is not subnormal (see, Halmos [1,p.107]).

As specific examples of subnormal operators we mention the following.

1°. Let ℓ_2 and ℓ_2^+ be the Hilbert spaces of square summable complex sequences of the form $\{c_n\}_{n=-\infty}^{\infty}$ and $\{c_n\}_{n=0}^{\infty}$, respectively. Let $\mathcal{M}$ be an auxiliary Hilbert space and write

$$\ell_2(\mathcal{M}) = \ell_2 \otimes \mathcal{M}, \quad \ell_2^+(\mathcal{M}) = \ell_2^+ \otimes \mathcal{M} .$$

In other words $\ell_2(\mathcal{M})$ ($\ell_2^+(\mathcal{M})$) consists of the $\mathcal{M}$-valued square summable sequences of the form $\{f_n\}_{n=-\infty}^{\infty}$ ($\{f_n\}_{n=0}^{\infty}$). The operator U on $\ell_2(\mathcal{M})$ defined by

$$U\{f_n\}_{n=-\infty}^{\infty} = \{f_{n-1}\}_{n=-\infty}^{\infty}$$

is referred to as a <u>bilateral shift</u> with multiplicity equal to the dimension of $\mathcal{M}$. The operator U is clearly unitary. The subspace

$\ell_2^+(\mathcal{H})$ (viewed naturally as a subspace of $\ell_2(\mathcal{H})$) is an invariant subspace for U. The subnormal operator $U_+ = U|\ell_2^+(\mathcal{H})$ is called a _vector-valued unilateral shift_. The range of the self-commutator $[U_+^*, U_+]$ consists of elements $\{f_n\}_{n=0}^{\infty} \in \ell_2^+(\mathcal{H})$, for which $f_n = 0$, when $n > 0$. It follows immediately, from Theorem 1.1 of Section 1, that the operator U_+ is pure.

2°. Let μ be a non-negative finite Borel measure in the complex plane having compact support K. The closure of the polynomials (in the variable z) in $L^2(d\mu)$ is denoted by $H^2(d\mu)$. The closure of the rational functions with poles off K is denoted by $R^2(d\mu)$. There holds

$$H^2(d\mu) \subset R^2(d\mu) \subset L^2(d\mu) ,$$

where we must admit the possible equality between any of these spaces. The subspaces $H^2(d\mu)$ and $R^2(d\mu)$ are obviously invariant under the normal operator M_z defined on $L^2(d\mu)$ by

$$M_z f(z) = zf(z) \quad .$$

The subnormal operators $M_z|_{H^2}$ and $M_z|_{R^2}$ have received considerable attention. A result of Bram [1] establishes that every subnormal operator with a cyclic vector is unitarily equivalent to an operator of the form $M_z|_{H^2(d\mu)}$, for some measure μ.

3°. In this example we use the notations $\ell_2(\mathcal{H})$ and $\ell_2^+(\mathcal{H})$ introduced in Example 1°. Let $\{A_n\}_{n=-\infty}^{\infty}$ be a sequence of operators on $\mathcal{H}$ satisfying

$$\|A_n\| \le M, \; n = 0, \pm 1, \pm 2, \; \dots ,$$

where $M > 0$ is a constant. The operator B on $\ell_2(\mathcal{H})$ defined by

$$B\{f_n\}_{n=-\infty}^{\infty} = \{A_n f_{n-1}\}_{n=-\infty}^{\infty}$$

is called an operator valued bilateral weighted shift. Similarly, if $\{A_n\}_{n=0}^{\infty}$ is a sequence of operators on $\mathcal{H}$ satisfying

$$\|A_n\| \leq M, \quad n = 0,1,2,\ldots,$$

then we define the operator valued unilateral weighted shift A on $\ell_2^+(\mathcal{H})$ by

$$A\{f_n\}_{n=0}^{\infty} = \{A_n f_{n-1}\}_{n=0}^{\infty} \quad (f_{-1} = 0).$$

The bilateral weighted shift B is hyponormal if and only if $A_n^* A_n \geq A_{n-1}^* A_{n-1}$, for all n. Similarly, the unilateral weighted shift A is hyponormal if and only if $A_n^* A_n \geq A_{n-1}^* A_{n-1}$, for all $n \geq 1$.

The question of purity of the weighted shift operators seems complicated. Let us consider a more specific example. Let V and D be non-negative self-adjoint operators on $\mathcal{H}$. We will assume the range of V is dense. Set $A_n = \sqrt{V+D}$, $n \geq 0$, and $A_n = \sqrt{V}$, $n < 0$. The bilateral shift B in this case has the matrix representation

$$B = \begin{bmatrix} & \vdots & \vdots & \vdots & \vdots & \\ \cdots & \sqrt{V} & 0 & 0 & 0 & \cdots \\ \cdots & 0 & \sqrt{V} & \boxed{0} & 0 & \cdots \\ \cdots & 0 & 0 & \sqrt{V+D} & 0 & \cdots \\ \cdots & 0 & 0 & 0 & \sqrt{V+D} & \cdots \\ & \vdots & \vdots & \vdots & \vdots & \end{bmatrix}$$

and the self-commutator of B has the form

$$[B^*,B] = \begin{bmatrix} & \vdots & \vdots & \vdots & \\ \cdots & 0 & 0 & 0 & \cdots \\ \cdots & 0 & \boxed{D} & 0 & \cdots \\ \cdots & 0 & 0 & 0 & \cdots \\ & \vdots & \vdots & \vdots & \end{bmatrix} .$$

The operator B is pure if and only if the smallest subspace of $\ell_2(\mathcal{H})$ reducing B and containing the range of $[B^*,B]$ is all of $\ell_2(\mathcal{H})$ (see, Theorem 1.1). This happens if and only if $\mathcal{H}[V,D] = \mathcal{H}$, where $\mathcal{H}[V,D]$ denotes the smallest subspace of $\mathcal{H}$ reducing V containing the range of D.

4°. This example involves the Hilbert transform on the real line $\mathbb{R}$. Let f be a function in the Hilbert space $L^2(\mathbb{R})$. The <u>Hilbert transform</u> Qf of the function $f \in L^2(\mathbb{R})$ is defined a.e. by the equation

$$Qf(x) = \frac{1}{\pi i}\int_{\mathbb{R}} \frac{f(t)}{t-x}\, dt \equiv \lim_{\varepsilon\to 0} \frac{1}{\pi i}\int_{|t-x|\geq\varepsilon} \frac{f(t)}{t-x}\, dt. \qquad (2.1)$$

The singular integral defined in (2.1) is referred to as a Cauchy principal value. It requires a non-trivial argument to establish the existence a.e. of the Cauchy principal value integral in (2.1). Further, the operator $f \to Qf$ is a bounded self-adjoint operator on $L^2(\mathbb{R})$. The interested reader can find a pleasant real variable proof of these facts in the book of Garsia [1]. A more formal complex variable treatment of the Hilbert transform may be found in the book of Titchmarsh [1]. Later, in Section 3 of Chapter 2, we will have a little more to say about the Fourier transform of the operator Q. Here we wish only to make use of formal algebraic properties of Q.

Let $E \subset \mathbb{R}$ be a bounded, measurable subset. Assume $a, b \in L^\infty(E)$ with a real valued a.e. We define the self-adjoint operator Y on $L^2(E)$ by

$$Yf(x) = [af - bQ_E\bar{b}]f(x) = a(x)f(x) + \frac{b(x)}{\pi i}\int_E \frac{\overline{b(t)}\, f(t)}{x-t}\, dt,$$

where we have employed the notation Q_E for the compression of the operator Q to $L^2(E)$. The operator X is defined on $L^2(E)$ by the identity

$$Xf(x) = xf(x).$$

Write $H = X+iY$, so that,

$$Hf(x) = xf(x) + i\left[a(x)f(x) + \frac{b(x)}{\pi i}\int_E \frac{\overline{b(t)}f(t)}{x-t}\,dt\right]. \quad (2.2)$$

A formal calculation, using the identity $[H^*,H] = 2i[XY-YX]$ (cf. (1.3)) and the fact that multiplication by a commutes with X, leads to the identity

$$[H^*,H]f(x) = \frac{2}{\pi}\, b(x)\int_E f(t)\overline{b(t)}\,dt = \frac{2}{\pi}(f,b)b(x).$$

In other words, the operator H defined by (2.2) is a hyponormal operator on $L^2(E)$ with a one dimensional self-commutator.

This last example can be considerably generalized. One of the main results, in Chapter 2 of these notes, provides a singular integral representation, analogous to the representation (2.2), for an arbitrary hyponormal operator. Unfortunately, this representation does not resolve the problem of determining the structure (for example, invariant subspaces) of seminormal operators.

We conclude this section with a discussion of the purity of the operator $H = X+iY$ defined by (2.2) on $L^2(E)$. It is clear that if b is zero on some set $F \subset E$ having positive measure, then $L^2(F)$ is a reducing subspace for the operator H. Restricted to $L^2(F)$, the operator H is the normal operator

$$Hf(x) = xf(x) + i\,a(x)f(x) .$$

On the other hand, the smallest reducing subspace for the operator H containing the range of $[H^*,H]$ contains the closed linear manifold spanned by $\{X^n b\}_{n=0}^{\infty}$. If $b(t) \neq 0$ a.e., then this latter collection of functions span $L^2(E)$. The conclusion is that H is pure if and only if $b(t) \neq 0$ almost everywhere.

3. Local Spectral Theory for Hyponormal Operators

Let T be in $L(\mathcal{H})$. The resolvent set of the operator T will be denoted by $\rho(T) = \mathbb{C}\backslash\sigma(T)$. For a fixed vector $f \in \mathcal{H}$, the $\mathcal{H}$-valued analytic function

$$f(\lambda) = (T-\lambda)^{-1}f, \ \lambda \in \rho(T)$$

is called the local resolvent of the operator T for the vector f. It may be that the local resolvent $f(\lambda)$ possesses analytic continuations onto portions of $\sigma(T)$ and, in general, there may be many different extensions of the local resolvent onto portions of $\sigma(T)$. To avoid the latter difficulty we make the following definition. An operator $T \in L(\mathcal{H})$ is said to possess the single valued extension property in case the only $\mathcal{H}$-valued analytic function g satisfying $(T-\lambda)g(\lambda) = 0$, on any open set in the plane, is the function $g(\lambda) \equiv 0$. Let T be an operator with the single valued extension property. Then for f in $\mathcal{H}$, the local resolvent $f(\lambda)$ admits a maximal single valued analytic continuation to some open set containing $\rho(T)$. The domain of this maximal continuation of the local resolvent $(T-\lambda)^{-1}f$ will be denoted by $\rho_T(f)$. The set $\rho_T(f)$ is referred to as the local resolvent set and the complementary set $\sigma_T(f) = \mathbb{C}\backslash\rho_T(f)$ is called the local spectrum of the vector f.

One way to guarantee that the operator T have the single valued extension property is to assume that the point spectrum (the set of eigenvalues) $\pi_0(T)$ is empty. Later, at several stages, this assumption will be made. Fortunately, in the case of hyponormal operators, this hypothesis is superfluous.

Lemma 3.1. Let H be hyponormal operator on $\mathcal{H}$, then H has the single valued extension property.

Proof. Suppose first that N is a normal operator on $\mathcal{H}$ and, for some non-empty, open subset Ω of the plane, there is an analytic function $z(\lambda)$ satisfying $(N-\lambda)z(\lambda) = 0$, $\lambda \in \Omega$. The vector $z(\lambda)$ is an eigenvector of N corresponding to λ. Since eigenvectors of a normal operator corresponding to distinct eigenvalues are orthogonal, then for $\lambda \neq \lambda_0 \in \Omega$,

$$\|z(\lambda) - z(\lambda_0)\|^2 = \|z(\lambda)\|^2 + \|z(\lambda_0)\|^2 \quad .$$

Unless $z(\lambda_0) = 0$, this last identity precludes even (strong) continuity of the function $z(\lambda)$ at λ_0. The conclusion is that $z(\lambda) \equiv 0$, $\lambda \in \Omega$.

For the case of a hyponormal operator H, we can appeal to Theorem 1.1 and write $H = H_0 \oplus N$, where H_0 is pure and N is normal. The purity of H_0 implies $\pi_0(H_0) = \phi$. Thus both H_0 and N have the single valued extension property and this completes the proof of the lemma.

We remark that Lemma 3.1 is not true for cohyponormal operators. In fact, let U_+ be the unilateral shift on ℓ_2^+. Set $f = e_0 = \{\delta_{0j}\}_{j=0}^{\infty}$. In the punctured disc $\Delta' = \{\lambda : 0<|\lambda|<1\}$, both of the functions

$$f_1(\lambda) = -\frac{1}{\lambda} e_0, \; f_2(\lambda) = \sum_{j=0}^{\infty} \lambda^j U_+^{j+1} e_0$$

are analytic ℓ_2^+-valued functions. For $i = 1,2$,

$$(U_+^* - \lambda) f_i(\lambda) = f, \quad \text{for all } \lambda \in \Delta' \; .$$

This last example (due to Stampfli [3]) shows that the cohyponormal operator U_+^* does not have the single valued extension property.

For δ a closed subset of the plane, we define the <u>local spectral subspace</u> $\mathcal{M}_T(\delta)$ as the set

$$\mathcal{M}_T(\delta) = \{f \in \mathcal{H}: \sigma_T(f) \subset \delta\}.$$

It is an easy exercise to verify that $\mathcal{M}_T(\delta)$ is a linear manifold invariant under the operator T. It develops that $\mathcal{M}_T(\delta)$ is not always closed; however, as we shall presently see, when the operator T is hyponormal, then the manifold $\mathcal{M}_T(\delta)$ is a closed subspace of $\mathcal{H}$ invariant under the operator T.

We deduce a pair of simple lemmas concerned with the local resolvent:

Lemma 3.2. Let T be an operator on $\mathcal{H}$ with the single valued extension property. Let f be in $\mathcal{H}$. Denote by $f(\lambda)$ the local resolvent of the vector f defined on $\rho_T(f)$. Then, for $\lambda \in \rho_T(f)$,

$$(T-\lambda)^{n+1} f^{(n)}(\lambda) = n!f, \quad n = 0,1,2,\ldots . \tag{3.1}$$

In particular, the vector f is in the range of $(T-\lambda)^n$, for $n = 0,1,2,\ldots$, and if the operator $T-\lambda$ is one-to-one, then

$$(T-\lambda)^{-n} f = \frac{1}{(n-1)!} f^{(n-1)}(\lambda), \quad n = 1,2,\ldots . \tag{3.2}$$

Proof. The identity (3.1) is established by induction. The result is obviously valid when $n = 0$. If (3.1) holds for the integer $p-1$, then differentiating, we obtain $(T-\lambda)^p f^{(p)}(\lambda) = p(T-\lambda)^{p-1} f^{(p-1)}(\lambda)$. Operating on both sides of this last identity by $T-\lambda$ and taking advantage of the inductive hypothesis we obtain (3.1) for $n=p$. The remaining statements of the lemma are now obvious. This completes the proof.

We will introduce the following notation. Let $\gamma: [0,1] \to \mathbb{C}$ be a simple closed rectifiable curve. For $\lambda \notin \gamma$, the index of the point with respect to γ is defined as the integer $\operatorname{ind}_\gamma(\lambda)$ equal to $(2\pi)^{-1}$ times the jump in the argument of the function $\gamma-\lambda$ on the interval $[0,1]$. For the case where Γ is a finite union $\Gamma = \gamma_1 \cup \ldots \cup \gamma_N$ of non-intersecting simple closed curves, we define

$$\operatorname{ind}_\Gamma(\lambda) = \sum_{i=1}^{N} \operatorname{ind} \gamma_i(\lambda), \quad \lambda \notin \Gamma.$$

Lemma 3.3. Let T be an operator on $\mathscr{H}$ without point spectrum. Let $f \in \mathscr{H}$, and let Ω be an open set containing $\sigma_T(f)$. Assume Γ is a subset of $\Omega \setminus \sigma_T(f)$, which is a finite disjoint union of simple closed rectifiable curves, with the property that

$$\mathrm{ind}_\Gamma(\lambda) = \begin{cases} 1 & \lambda \in \sigma_T(f) \\ 0 & \lambda \notin \Omega . \end{cases} \tag{3.3}$$

Then for $n = 0, \pm 1, \pm 2, \ldots,$

$$(T-\lambda_0)^n f = -\frac{1}{2\pi i}\int_\Gamma (\lambda - \lambda_0)^n f(\lambda)\,d\lambda, \quad \lambda_0 \notin \Omega. \tag{3.4}$$

Proof. First we establish (3.4) for the case where $n \geq 0$. When $R > \|T\|$,

$$-\frac{1}{2\pi i}\int_\Gamma (\lambda-\lambda_0)^n f(\lambda)\,d\lambda = -\frac{1}{2\pi i}\int_{|\lambda|=R} (\lambda-\lambda_0)^n f(\lambda)\,d\lambda. \tag{3.5}$$

This follows from Cauchy's Theorem. The right hand side of (3.5) equals $(T-\lambda_0)^n f$. Thus (3.4) holds when $n \geq 0$.

Next let $m \geq 1$ be fixed. Set

$$\beta = -\frac{1}{2\pi i}\int_\Gamma (\lambda-\lambda_0)^{-m} f(\lambda)\,d\lambda .$$

Then

$$(T-\lambda_0)^m \beta = \frac{1}{2\pi i}\int_\Gamma [(T-\lambda)-(\lambda-\lambda_0)]^m (\lambda - \lambda_0)^{-m} f(\lambda)\,d\lambda$$

$$= -\sum_{j=0}^{m} \frac{1}{2\pi i}\int_\Gamma \binom{m}{j} (\lambda-\lambda_0)^{-j} (T-\lambda)^j f(\lambda)\,d\lambda .$$

For $j > 0$ and $\lambda_0 \notin \Omega$, we have $\int_\Gamma (\lambda-\lambda_0)^{-j}(T-\lambda)^j f(\lambda)\,d\lambda = 0$. This last statement follows from the fact that $(T-\lambda)^j f(\lambda) = (T-\lambda)^{j-1} f$ and Cauchy's Theorem. Thus $(T-\lambda_0)^m \beta = f$. This establishes (3.4) for negative n and completes the proof of the lemma.

The next lemma is special to hyponormal operators:

Lemma 3.4. Let H be a hyponormal operator without eigenvalues and f a unit vector in $\mathcal{H}$. If $\lambda_0 \in \rho_H(f)$, then

$$\|(H-\lambda_0)^{-1}f\|^n \leq \|(H-\lambda_0)^{-n}f\|, \; n = 1,2,\ldots \quad . \tag{3.6}$$

Proof. The existence of the vectors $(H-\lambda_0)^{-n}f$, $n=1,2,\ldots$, follows from (3.2). Set $a_n = \|(H-\lambda_0)^{-n}f\|$, $n = 1,2,\ldots$. Then

$$\begin{aligned} a_1^2 = \|(H-\lambda_0)^{-1}f\|^2 &= ((H-\lambda_0)^{-1}f,(H-\lambda_0)^{-1}f) \\ &= ((H-\lambda_0)(H-\lambda_0)^{-2}f,(H-\lambda_0)^{-1}f) \\ &\leq \|(H-\lambda_0)^{-2}f\| \, \|(H-\lambda_0)^*(H-\lambda_0)^{-1}f\| \\ &\leq \|(H-\lambda_0)^{-2}f\| = a_2 \, . \end{aligned}$$

Similarily, $a_n^2 \leq a_{n+1}a_{n-1}$, $n = 2,3,\ldots$. The inequality (3.6) follows directly from Lemma 1.2. This completes the proof.

We can now prove Stampfli's local analogue of Corollary 1.1:

Theorem 3.1. Let H be a hyponormal operator without eigenvalues and f a unit vector in $\mathcal{H}$. Then for $\lambda \in \rho_H(f)$,

$$\|f(\lambda)\| \leq \frac{1}{\operatorname{dist}(\lambda,\rho_H(f))} \quad . \tag{3.7}$$

Proof. Let $\lambda_0 \in \rho_H(f)$. For any preassigned $\varepsilon > 0$, choose an open set Ω containing $\sigma_H(f)$ such that $\operatorname{dist}(\sigma_H(f), \mathbb{C} \setminus \Omega) < \varepsilon$. Let $\subset \Omega \setminus \sigma_H(f)$ be a system of simple closed rectifiable curves having property (3.3) and assume $\lambda_0 \notin \Omega$. Making use of Lemmas 3.3 and 3.4, we obtain the following estimates :

$$\|f(\lambda_0)\| = \|(H-\lambda_0)^{-1}f\|$$
$$\leq \|(H-\lambda_0)^{-n}f\|^{1/n}$$
$$= \left\|\frac{1}{2\pi i}\int_\Gamma (\lambda-\lambda_0)^{-n}f(\lambda)d\lambda\right\|^{1/n}$$
$$\leq \frac{[M|\Gamma|]^{1/n}}{(2\pi)^{1/n}\mathrm{dist}(\lambda_0,\Gamma)},$$

where $|\Gamma|$ denotes the length of Γ and $M = \max\{\|f(\lambda)\|:\lambda \in \Gamma\}$. Letting $n \to \infty$, we obtain for small $\varepsilon > 0$,

$$\|f(\lambda_0)\| \leq \frac{1}{\mathrm{dist}(\lambda_0,\Gamma)} \leq \frac{1}{\mathrm{dist}(\lambda_0,\sigma_H(f))-\varepsilon} .$$

As $\varepsilon > 0$ can be chosen arbitrarily close to zero, this gives the desired estimate. The proof of the theorem is complete.

The next result of Stampfli [8] is the most important result concerning the local spectrum of a hyponormal operator:

Theorem 3.2. Let H be a hyponormal operator on $\mathcal{H}$ without eigenvalues. For any closed subset δ of the plane, the linear manifold $\mathcal{M}_H(\delta)$ is a closed invariant subspace for the operator H.

Proof. Let $\{f_n\}_{n=1}^\infty$ be a sequence of vectors in $\mathcal{M}_H(\delta)$ having limit f. Let Δ be a closed disc contained in $\mathbb{C}\backslash\delta$. Directly from (3.7) of Theorem 3.1, we conclude that there is a constant $M > 0$ such that, for $\lambda \in \Delta$ and $n = 1,2,...,$

$$\|f_n(\lambda)\| \leq M .$$

A normal family argument can be used to establish that there is a subsequence $\{f_{n_k}\}_{k=0}^\infty$ such that $f(\lambda) = \lim_k f_{n_k}(\lambda)$ defines a (bounded) analytic function on the interior of Δ. Obviously,

$(H-\lambda)f(\lambda) = \lim_k (H-\lambda)f_{n_k}(\lambda) = \lim_k f_{n_k} = f$. This implies $\sigma_H(f) \subset \delta$ and completes the proof of the theorem.

It should be noted that Theorem 3.2 remains valid in case the hypothesis concerning the eigenvalues of H is dropped. To see this we require the following lemma adapted from Putnam [6]:

Lemma 3.5. Let N be a normal operator on $\mathcal{H}$. Assume $N = \int z dG_z$ is the spectral resolution of the operator N. If β is any Borel set and $f \in \bigcap_{\lambda \in \beta} R(N-\lambda)$, then $G(\beta)f = 0$.

Proof. Let λ be a fixed complex number. For $s > 0$ we denote by $\Delta(\lambda:s)$ the open disc of radius s centered at λ. It will first be shown that whenever $f \in R(N-\lambda)$, then

$$\overline{\lim_{s\to 0^+}} [s^{-2} \int_{\Delta(\lambda:s)} d\|G_z f\|^2] = 0 \quad . \tag{3.8}$$

Let $f(\lambda)$ denote any solution of $(N-\lambda)f(\lambda) = f$ and let $g(\lambda)$ denote its orthogonal projection onto $[\ker(N-\lambda)]^{\perp}(=[\ker(N-\lambda)^*]^{\perp})$. Then

$$s^{-2}\int_{\Delta(\lambda:s)} d\|G_z f\|^2 \leq \int_{\Delta(\lambda:s)} \frac{1}{|z-\lambda|^2} d\|G_z f\|^2 = \int_{\Delta(\lambda:s)} d\|G_z g(\lambda)\|^2 \quad .$$

Consequently, equation (3.8) holds, whenever $f \in R(N-\lambda)$.

Now if β is any Borel set and $f \in \bigcap_{\lambda \in \beta} R(N-\lambda)$, then (3.8) holds at every $\lambda \in \beta$. This implies $\|G(\beta)f\| = 0$ (see, e.g. Saks [1, §14-15]). The proof of the lemma is complete.

Theorem 3.3. Let N be a normal operator on $\mathcal{H}$ and assume $N = \int z dG_z$ is the spectral resolution of N. Let δ be a closed set in the plane. Then

$$\mathcal{M}_N(\delta) = \bigcap_{\lambda \notin \delta} R(N-\lambda) = G(\delta)\mathcal{H} \quad .$$

In particular, $\mathcal{M}_N(\delta)$ is a closed subspace of $\mathcal{H}$.

Proof. The inclusion $\mathcal{M}_N(\delta) \subset \bigcap_{\lambda \notin \delta} R(N-\lambda)$ is obvious. The inclusion $\bigcap_{\lambda \notin \delta} R(N-\lambda) \subset G(\delta)\mathcal{H}$ is a reformulation of Lemma 3.5 for the case $\beta = \mathbb{C}\setminus\delta$. There remains to show $G(\delta)\mathcal{H} \subset \mathcal{M}_N(\delta)$. To this end, suppose $f \in G(\delta)\mathcal{H}$. For $\lambda \notin \delta$, define

$$f(\lambda) = \frac{1}{2\pi i}\int_\delta \frac{1}{z-\lambda}\, dG_z f \quad .$$

The function $f(\lambda)$ is an analytic $\mathcal{H}$-valued function on $\mathbb{C}\setminus\delta$ which satisfies $(N-\lambda)f(\lambda) = f$. This means $\sigma_N(f) \subset \delta$ or, equivalently, $f \in \mathcal{M}_N(\delta)$. This completes the proof of Theorem 3.3.

Corollary 3.1. Let H be hyponormal operator on $\mathcal{H}$. For any closed subset δ of the plane, $\mathcal{M}_H(\delta)$ is a closed invariant subspace for H.

This last corollary implies the following:

Corollary 3.2. Let H be a hyponormal operator on the Hilbert space $\mathcal{H}$. Suppose there is a single non-zero vector $f \in \mathcal{H}$ such that $\sigma_H(f)$ is a proper subset of $\sigma(H)$. Then the operator H has a non-trivial invariant subspace.

Unfortunately, this last corollary does not provide invariant subspaces for an arbitrary hyponormal operator. This can be seen from the following simple example:

First note, if T is an operator with the single valued extension property and $\mathcal{M} = \mathcal{M}_T(\delta)$ is a local spectral space which is closed, then $\sigma(T|_{\mathcal{M}}) \subset \delta$.

Consider the operator U_+ on ℓ_2^+. If $\mathcal{M}$ is a closed non-zero U_+ - invariant subspace, then

$$\sigma(U_+|\mathcal{M}) = \{z: |z| \leq 1\} \quad .$$

This last remark follows from the pure isometric nature of U_+. Consequently, there are no vectors f in ℓ_2^+ such that $\sigma_{U_+}(f)$ is a proper subset of $\sigma(U_+)$.

Later in Section 5, we will use Corollary 3.2 to obtain invariant subspaces for a few hyponormal operators.

In Theorem 3.3, it was established that

$$\mathcal{M}_N(\delta) = \bigcap_{\lambda \notin \delta} R(N-\lambda) , \qquad (3.9)$$

when N is a normal operator. The remainder of this section is devoted to generalizing (3.9) to the case of hyponormal operators. Along the way we derive some results of independent interest.

Let A be an operator on $\mathcal{H}$. By <u>the polar factorization</u> of A we mean the representation A=WR, where $R=(A^*A)^{1/2}$ and the operator W is the unique partial isometry which is zero on ker A, having final space the closure of range of A, satisfying A = WR.

<u>Lemma 3.6.</u> Let $T \in L(\mathcal{H})$. Assume $T^* = WR$ is the polar factorization of the operator T^*. Let $R = \int_0^\infty t\,dE(t)$ be the spectral resolution of R. Suppose f,g satisfy Tg = f, then

$$\|Pg\|^2 = \int_0^\infty t^{-2} d\|E(t)f\|^2 = \lim_{s\to 0^+} ((R+s)^{-2}f,f), \qquad (3.10)$$

where P denotes the orthogonal projection onto $[\ker T]^\perp$.

Proof. For s > 0,

$$((R+s)^{-2}f,f) = \int_0^\infty (t+s)^{-2} d\|E(t)f\|^2 .$$

An application of the monotone convergence theorem establishes the second equality in (3.10). This last argument uses the fact that $E(\{0\})f = 0$. From the identity $RW^*Pg = f$, we obtain

$$\|W^*Pg\|^2 = \int_0^\infty t^{-2} d\|E(t)f\|^2 .$$

It is easy to check that $\|W^*Pg\| = \|Pg\|$. This completes the proof of the lemma.

<u>Lemma 3.7.</u> Let T be in $L(\mathcal{H})$. Fix $f \in \mathcal{H}$ and denote by $Z = Z_f$ the set of complex numbers λ such that f belongs to $R(T-\lambda)$. For $\lambda \in Z$, let $f_0(\lambda)$ be the unique solution of $(T-\lambda)x = f$ which is in $[\ker(T-\lambda)]^{\perp}$. Then $\|f_0(\lambda)\|$ is a lower semi-continuous function on Z.

Proof. Let $\lambda \in Z$ and let $T_\lambda^* = W(\lambda)R(\lambda)$ be the polar factorization of T_λ^*. From Lemma 3.6

$$\|f_0(\lambda)\|^2 = \lim_{s\to 0^+} ((R(\lambda)+s)^{-2}f,f) \quad , \ \lambda \in Z.$$

For $s > 0$ fixed, the function $((R(\lambda)+s)^{-2}f,f)$ is continuous on Z. Thus $\|f_0(\lambda)\|$ is the supremum of a collection of continuous functions and is, therefore, lower semi-continuous on Z. This completes the proof.

Let T be an operator on $\mathcal{H}$ and f a fixed vector in $\mathcal{H}$. Suppose on some set Z in the plane there is a bounded $\mathcal{H}$-valued function $f(\lambda)$ satisfying $(T-\lambda)f(\lambda) = f$, $\lambda \in Z$. If no point in Z is an eigenvalue of the operator T, then the vector function $\lambda \to f(\lambda)$ is weakly continuous on Z. The proof of this remark is easy and is left as an exercise.

For the case where Z is open and T is hyponormal, the following theorem of Radjabalipour [1] significantly strengthens the assertion in the last paragraph. The clever proof of this theorem is from Stampfli and Wadhwa [1]:

<u>Theorem 3.4.</u> Let H be hyponormal operator on $\mathcal{H}$ and f a vector in $\mathcal{H}$. Suppose the disc Δ contains no eigenvalues of H and that there is a bounded function $f(\lambda)$ on Δ satisfying $(H-\lambda)f(\lambda) = f$, $\lambda \in \Delta$. Then $f(\lambda)$ is is analytic on Δ.

Proof. By the remark above, the function $f(\lambda)$ is weakly continuous on Δ.

Suppose the vector g is in $R(H-\lambda_0)$, for some $\lambda_0 \in \Delta$. Taking advantage of the fact that $R(H-\lambda_0) \subset R[(H-\lambda_0)^*]$ (see, Remark 2° in Section 1), we obtain a vector h_0 such that $(H-\lambda_0)^* h_0 = g$. In this case, the function $(f(\lambda),g)$ is differentiable at λ_0. In fact,

$$\lim_{\lambda\to\lambda_0} \frac{(f(\lambda),g) - (f(\lambda_0),g)}{\lambda - \lambda_0} = \lim_{\lambda\to\lambda_0} \frac{(H_{\lambda_0} f(\lambda),h_0) - (H_{\lambda_0} f(\lambda_0),h_0)}{\lambda - \lambda_0}$$

$$= \lim_{\lambda\to\lambda_0} (f(\lambda),h_0)$$

$$= (f(\lambda_0),h_0) \quad .$$

Now let Γ be a triangle contained in Δ. Set

$$g = \frac{1}{2\pi i} \int_\Gamma f(\lambda)d\lambda \quad ,$$

here this integral is considered as a weak integral. For λ_0 in the nterior of Γ, the vector

$$g(\lambda_0) = \frac{1}{2\pi i} \int_\Gamma \frac{f(\lambda)}{\lambda-\lambda_0} d\lambda - f$$

atisfies $(H-\lambda_0)g(\lambda_0) = g$. Consequently, the function $(f(\lambda),g)$ is nalytic on the interior of the triangle Γ and continuous on Γ. Thereore,

$$\|g\|^2 = \frac{1}{2\pi i} (\int_\Gamma f(\lambda)d\lambda,g) = \frac{1}{2\pi i} \int_\Gamma (f(\lambda),g)d\lambda = 0.$$

his implies $\int_\Gamma f(\lambda)d\lambda = 0$, for every triangle Γ contained in Δ, and stablishes the analyticity of the function $f(\lambda)$. The proof of the emma is complete.

We are now in a position to prove the generalization of (3.9) or hyponormal operators.

Theorem 3.5. Let H be a hyponormal operator on $\mathcal{H}$ and let δ be a closed subset of the plane. Then

$$\mathcal{M}_H(\delta) = \bigcap_{\lambda \notin \delta} R(H-\lambda) . \tag{3.11}$$

Proof. First we apply the result of Theorem 1.1 to write $H = H_0 \oplus N$, where H_0 is a pure hyponormal operator and N is normal. The conclusion of the theorem for the normal part was obtained in Theorem 3.3. It suffices to establish (3.11) when the operator H is pure. This ensures that H has no eigenvalues.

The inclusion $\mathcal{M}_H(\delta) \subset \bigcap_{\lambda \notin \delta} R(H-\lambda)$ is obvious. We will establish the reverse inclusion. Let f be a vector in $\bigcap_{\lambda \notin \delta} R(H-\lambda)$. Denote by $f(\lambda)$ the unique vector satisfying $(H-\lambda)f(\lambda) = f$, $\lambda \notin \delta$. It will be shown that the function $f(\lambda)$ is analytic on $\mathbb{C}\backslash\delta$.

Let Δ be a closed disc contained in $\mathbb{C}\backslash\delta$ and Δ^0 its interior. Suppose that $\Delta^0 \cap \sigma_H(f) \neq \emptyset$. For $n = 1,2,\ldots$, let

$$F_n = \{\lambda \in \Delta \cap \sigma_H(f) : \|f(\lambda)\| \leq n\} .$$

The lower semicontinuity of $\|f(\lambda)\|$ (see, Lemma 3.7) implies that each F_n is closed. Since $\bigcup_{n=1}^{\infty} F_n = \Delta \cap \sigma_H(f)$, then the Baire Category Theorem implies, for some m, the set $F_m \cap \Delta^0$ has interior in the relative topology on $\Delta \cap \sigma_H(f)$.

This means there is an open disc $\Delta_1 \subset \Delta^0$ with center in $\sigma_H(f)$ such that $\|f(\lambda)\| \leq m$, for all $\lambda \in \Delta_1 \cap \sigma_H(f)$. Let Δ_2 be the disc with same center as Δ_1 and radius equal to one-half of the radius of Δ_1. Let $\lambda_0 \in \Delta_2 \cap \rho_H(f)$ (if such a λ_0 exists) and choose γ_0 in $\Delta_1 \cap \sigma_H(f)$ such that $|\lambda_0-\gamma_0| = \operatorname{dist}(\lambda_0,\sigma_H(f))$. Set $g_0 = f(\gamma_0) = H_{\gamma_0}^{-1}f$. It is easy to check that $\sigma_H(f) = \sigma_H(g_0)$. Further, for $\lambda \in \rho_H(f)$,

$$g(\lambda) = \frac{f(\lambda) - f(\gamma_0)}{\lambda - \gamma_0}$$

satisfies $(H-\lambda)g(\lambda) = g_0$. It follows from (3.7) that

$$\|g(\lambda_0)\| = \frac{\|f(\lambda_0)-f(\gamma_0)\|}{|\lambda_0 - \gamma_0|} \leq \frac{\|f(\gamma_0)\|}{\operatorname{dist}(\lambda_0,\sigma_H(g_0))} .$$

Since $\operatorname{dist}(\lambda_0,\sigma_H(g_0)) = \operatorname{dist}(\lambda_0,\sigma_H(f)) = |\lambda_0-\gamma_0|$, then

$$\|f(\lambda_0)\| \leq 2\|f(\gamma_0)\| \leq 2m .$$

In any case we conclude $\|f(\lambda)\| \leq 2m$, $\lambda \in \Delta_2$. By Lemma 3.8, $f(\lambda)$ is analytic on Δ_2. This contradicts the assumption that the center of Δ_2 is in $\sigma_H(f)$. The proof of the theorem is complete.

Corollary 3.3. Let H be a hyponormal operator. Then

$$\bigcap_{\lambda \in \mathbb{C}} R(H-\lambda) = \{0\} .$$

Proof. Without loss of generality we can assume that the operator H is pure. This reduction is accomplished by appealing to Theorem 1.1 and Theorem 3.3. This allows us to assume $\pi_0(H) = \emptyset$.

Suppose $f \in \bigcap_{\lambda \in \mathbb{C}} R(H-\lambda)$. Then there is a unique solution $f(\lambda)$ of $(H-\lambda)f(\lambda) = f$, for all $\lambda \in \mathbb{C}$. Directly from Theorem 3.5, it follows that $f(\lambda)$ is an $\mathcal{H}$-valued entire function. Since $f(\lambda) = H_\lambda^{-1}f$ vanishes at infinity, then $f(\lambda) \equiv 0$. Therefore, $f = 0$. The proof of the corollary is complete.

It is easy to see that Corollary 3.3 does not hold for arbitrary operators (consider the case of a nilpotent). In the next section, we will see that Corollary 3.3 is always false for non-normal cohyponormal operators.

4. Local Spectral Theory for Cohyponormal Operators

In this section we will present some results on the local spectral theory of cohyponormal operators. These results stand in contrast to the results for the hyponormal case obtained in the preceeding section.

We begin with the following simple lemma:

Lemma 4.1. Let A,B be self-adjoint operators. Assume for some $\varepsilon > 0$, $0 \leq \varepsilon I \leq A \leq B$. Then $0 \leq B^{-1} \leq A^{-1}$.

Proof. Let $A^{1/2}$ denote the unique non-negative square root of the operator A. From the identity $A^{1/2}A^{1/2} \leq B$, we obtain $I \leq A^{-1/2}BA^{-1/2}$. Then $A^{1/2}B^{-1}A^{1/2} \leq I$ and, therefore, $B^{-1} \leq A^{-1/2}A^{-1/2} = A^{-1}$. This completes the proof.

Let C be a cohyponormal operator. The notation $D = [C,C^*]$ will be employed for the self-commutator of the hyponormal operator C^*. For every complex number z, we have

$$(C-z)(C-z)^* \geq (C-z)(C-z)^* - (C-z)^*(C-z) = D \geq 0. \tag{4.1}$$

The next result is from Putnam [8, 10]:

Theorem 4.1. Let C be a pure cohyponormal operator on the Hilbert space $\mathcal{H}$. Let $D = [C,C^*]$ have the spectral resolution $D = \int_0^\infty t\,dG(t)$. Suppose $f \in \mathcal{H}$ satisfies

$$f \perp \ker D \quad \text{and} \quad k_f \equiv \int_{0^+}^\infty t^{-1} d\|G(t)f\|^2 < \infty. \tag{4.2}$$

Then $f \in \bigcap_{z \in \mathbb{C}} R(C-z)$. If f(z) denotes the unique solution of $(C-z)x=f$, which is orthogonal to ker(C-z), then

$$\|f(z)\| \leq k_f^{1/2} . \tag{4.3}$$

Proof. It is sufficient to establish that any f satisfying (4.2) belongs to the range of C and to establish (4.3) when z = 0. Let $CC^* = \int_0^\infty t\,dE(t)$ denote the spectral resolution of CC^*. The purity of

C yields that $\ker CC^* = \ker C^* = \{0\}$. Consequently, $E(\{0\}) = 0$.

For any $\varepsilon > 0$, we have (see equation (4.1))

$$0 \le \varepsilon I \le D+\varepsilon I \le CC^*+\varepsilon I .$$

Directly from Lemma 4.1,

$$(CC^* + \varepsilon I)^{-1} \le (D+\varepsilon I)^{-1} .$$

Let f satisfy (4.2). The last inequality implies

$$\begin{aligned}\int_0^\infty (t+\varepsilon)^{-1} d\|E(t)f\|^2 &= ((CC^*+\varepsilon I)^{-1}f,f)\\ &\le ((D+\varepsilon I)^{-1}f,f)\\ &= \int_0^\infty (t+\varepsilon)^{-1} d\|G(t)f\|^2\\ &\le \int_{0^+}^\infty t^{-1} d\|G(t)f\|^2\\ &= k_f .\end{aligned}$$

This implies, for any positive δ,

$$\int_\delta^\infty t^{-1} d\|E(t)f\|^2 = \lim_{\varepsilon\to 0} \int_\delta^\infty (t+\varepsilon)^{-1} d\|E(t)f\|^2 \le k_f .$$

Therefore, if we define the vector

$$f_\delta = \int_\delta^\infty t^{-1/2} dE(t)f ,$$

then f_δ has the following properties:

$$\|f_\delta\| \le k_f^{1/2}, \qquad (CC^*)^{1/2} f_\delta = E[\delta,\infty)f . \tag{4.4}$$

Let g be a weak limit of some subsequence of $\{f_\delta\}_{\delta>0}$. It follows, from (4.4) and the strong right continuity of the spectral resolution $\{E(t)\}$ at $t = 0$, that

$$\|g\| \le k_f^{1/2} , \qquad (CC^*)^{1/2} g = f . \tag{4.5}$$

Now let $C^* = W(CC^*)^{1/2}$ be the polar factorization of the operator C^*.

The operator W is isometric so that $W^*W = I$. Set $f(0) = Wg$. The vector $f(0)$ is orthogonal to ker C. From (4.5), $\|f(0)\| = \|g\| \leq k_f^{1/2}$ and $Cf(0) = (CC^*)^{1/2}W^*Wg = f$. This completes the proof of the theorem.

The following result should be contrasted with Corollary 3.3:

Corollary 4.1. Let C be a cohyponormal operator on the Hilbert space $\mathcal{H}$ and let $D = [C,C^*]$. Denote by $\mathcal{L}(D)$ the smallest linear manifold containing the range of D that is invariant under the operator C. Then

$$\mathcal{L}(D) \subset \bigcap_{z \in \mathbb{C}} R(C-z) . \tag{4.6}$$

Proof. Any vector in the range of D satisfies (4.2). Thus $R(D) \subset \bigcap_{z \in \mathbb{C}} R(C-z)$. Since the right hand side of this last inclusion is invariant under C, then (4.6) follows. This completes the proof of the corollary.

Remark. Let C be a cohyponormal operator and f a non-zero vector in the linear manifold $\mathcal{L}(D)$ introduced in Corollary 4.1. For $\lambda \in \mathbb{C}$, let $f(\lambda)$ denote the unique solution of $(C-\lambda)x = f$, which is orthogonal to $\ker(C-\lambda)$. Combining the results in Theorem 4.1 and Lemma 3.7, we learn that $\|f(\lambda)\|$ is a bounded lower semi-continuous function on all of $\mathbb{C}$. The boundedness of the function $\|f(\lambda)\|$ implies, when $\pi_0(C) = \emptyset$, the weak continuity of the mapping $\lambda \to f(\lambda)$. It is easy to see that the mapping $\lambda \to f(\lambda)$ is not analytic (see the proof of Corollary 3.3). There are examples (which will not be discussed here) with $\pi_0(C) = \emptyset$ and such that $\lambda \to f(\lambda)$ is not even strongly continuous on $\mathbb{C}$.

In Section 3 an example of a cohyponormal operator C without the single valued extension property was described. On the other hand, when a cohyponormal operator has the single valued extension property, we have the following:

Theorem 4.2. Let C be a pure cohyponormal operator on $\mathcal{H}$ which possesses the single valued extension property. There is a non-zero vector $f \in \mathcal{H}$ with $\sigma_C(f)$ a proper subset of $\sigma(C)$.

Proof. We dispense of the case where C has eigenvalues. If $\lambda_0 \in \pi_0(C)$ and f_0 is non-zero with $(C-\lambda_0)f_0 = 0$, then it is plain that $\sigma_C(f_0) = \{\lambda_0\}$. However, if $\sigma(C) = \{\lambda_0\}$, then applying Proposition 1.2 to the operator $C-\lambda_0$, we conclude $C = \lambda_0 I$. This is impossible. Consequently, $\sigma_C(f_0) \subsetneq \sigma(C)$. This completes the proof when C has eigenvalues.

Assume $\pi_0(C) = \emptyset$. Let f be an element in the range of $[C,C^*]$. Theorem 4.1 ensures the existence of a bounded function $f(\lambda)$ satisfying $(C-\lambda)f(\lambda) = f$. As noted above, the function $f(\lambda)$ is weakly continuous but cannot be analytic at all points of $\sigma(C)$. Consequently, there is a triangle Γ with the property that

$$g = \frac{1}{2\pi i} \int_\Gamma f(z)dz \neq 0$$

and such that $\sigma(C)$ intersects the exterior of Γ. The vector function

$$g(\lambda) = \frac{1}{2\pi i} \int_\Gamma (z-\lambda)^{-1} f(z)dz \ , \quad \lambda \notin \Gamma$$

is analytic off Γ and satisfies $(C-\lambda)g(\lambda) = g$, in the exterior of Γ. This shows $\sigma_C(g) \subsetneq \sigma(C)$ and completes the proof of the theorem.

. Applications of Local Spectral Theory.

In the final section of this chapter we will provide two examples of invariant subspaces for hyponormal operators that are obtained through the use of local spectral theory. The first example involves the singular integral operators discussed in Section 2. The second example is a very modest attempt to split off a normal reducing subspace for a hyponormal operator having a "thin portion" in its spectrum. Later, in Chapter 3, we will have use for this second application. Both examples can be considerably improved.

Before proceeding to the first example we will discuss a result of Putnam [3] concerning the relationship between the spectrum of a seminormal operator and the spectrum of its real and imaginary parts. Here we prove only one-half of the result. The proof is completed in Section 1 of Chapter 2. It is inserted to help clarify the spectral properties of the examples.

If T is an operator on $\mathcal{H}$, we recall that the <u>approximate point spectrum</u> $\pi(T)$ consists of all complex λ such that there is a sequence $\{f_n\}_{n=1}^{\infty}$ in $\mathcal{H}$ satisfying $\underline{\lim}\|f_n\| > 0$ and $\underline{\lim}\|(T-\lambda)f_n\| = 0$. It is well known (see, e.g. Halmos [1, p. 39]) that the boundary of $\sigma(T)$ is a subset of $\pi(T)$.

<u>Proposition 5.1</u>. Let $S = X + iY$ be a seminormal operator on $\mathcal{H}$ with X and Y its real and imaginary components, respectively. Then

$$\mathrm{proj}_X(\sigma(S)) = \sigma(X),\ \mathrm{proj}_Y(\sigma(S)) = \sigma(Y), \tag{5.1}$$

where proj_X (proj_Y) denotes the projection onto the x-axis (y-axis).

Proof. Without loss of generality it can be assumed that $S = H = X + iY$ is hyponormal. Since the operators H and $iH = -Y + iX$ are simultaneously hyponormal, then it is sufficient to establish the first identity in (5.1).

Write $D = [H^*,H] \geq 0$. Then for $z = x + iY \in \mathbb{C}$, we have (see equation (1.1))

$$H_z^*H_z = X_x^2 + Y_y^2 + \frac{1}{2} D . \tag{5.2}$$

If $z' = x + iy' \in \sigma(H)$, then there is a $z = x + iy$ in the boundary of $\sigma(H)$. Consequently, there is a sequence of unit vectors $\{f_n\}_{n=1}^{\infty}$ with $\|H_z f_n\| \to 0$. From (5.2)

$$\|H_z f_n\|^2 = (H_z^* H_z f_n, f_n) = \|X_n f_n\|^2 + \|Y_y f_n\|^2 + \frac{1}{2}(Df_n, f_n). \quad (5.3)$$

Each term on the right side of (5.3) is non-negative. In particular, it follows that $\lim_n \|X_x f_n\| = 0$. This implies $x \in \sigma(X)$ and we have established the inclusion $\text{proj}_x \sigma(H) \subset \sigma(X)$.

The reverse inclusion $\sigma(X) \subset \text{proj}_x \sigma(H)$ is established in Putnam [3]. In Section 1 of the next chapter, we will give a separate proof of this inclusion.

We turn to our first example and application of local spectral theory.

Let E be a bounded measurable subset of the real line. Let $a,b \in L^\infty(E)$ with $a(t)$ real valued and $b(t) \neq 0$ almost everywhere. We will consider the pure hyponormal singular integral operator H defined for $f \in L^2(E)$ by

$$Hf(x) = xf(x) + i[a(x)f(x) + \frac{b(x)}{\pi i}\int_E \frac{\overline{b(t)}f(t)}{x-t}\, dt]. \quad (5.4)$$

The operator H is the same operator that was defined in equation (2.2) and briefly studied in Section 2.

The real part of the operator H is the operator X defined on $L^2(E)$ by $Xf(x) = xf(x)$. From Proposition 5.1 we have $\text{proj}_x \sigma(H) = \sigma(X)$. The set $\sigma(X)$ is well known to consist of the collection of real numbers t_0 such that every neighborhood of t_0 intersects E in a set of positive measure. We will denote this set by $\overline{E}^e$ and refer to $\overline{E}^e$ as the <u>essential closure</u> of E.

Before stating a sufficient condition for the operator H defined in (5.4) to have an invariant subspace, we remark that, at the time of this writing, it is unknown whether every operator of the form (5.4) has an invariant subspace. The result which we derive here is the following:

Proposition 5.2. Let H denote the hyponormal operator defined in (5.4). Assume there is a real number μ_0 satisfying

$$\min \overline{E}^e < \mu_0 < \max \overline{E}^e \tag{5.5}$$

and

$$\int_E \frac{|b(t)|^2}{(t-\mu_0)^2}\, dt < \infty \quad . \tag{5.6}$$

Then the operator H has a non-trivial invariant subspace.

Proof. Without loss of generality we can assume $\pi_0(H^*) = \emptyset$. Let μ_0 satisfy the hypothesis of the proposition and let λ be a complex number of the form $\lambda = \mu_0 + i\nu$. For any measurable subset F of E, we will denote by P_F the orthogonal projection of $L^2(E)$ onto $L^2(F)$. From equations (1.2) and (2.3), we have

$$H_\lambda H_\lambda^* = X^2_{\mu_0} + Y^2_\nu - \frac{1}{\pi}\,(\ ,b)b \geq P_F[X^2_{\mu_0} - \frac{1}{\pi}\,(\ ,b)b]P_F. \tag{5.7}$$

We will show condition (5.6) implies, for δ sufficiently small, the set $F_\delta = (\mu_0-\delta,\mu_0+\delta) \cap E$ satisfies

$$R_\delta \equiv P_{F_\delta}[X^2_{\mu_0} - \frac{1}{\pi}\,(\ ,b)b]P_{F_\delta} \geq 0 \tag{5.8}$$

In fact, if the operator R_δ has the negative eigenvalue $-\gamma^2$, $(\gamma>0)$, then for some non-zero $f \in L^2(F_\delta)$

$$X^2_{\mu_0} f - \frac{1}{\pi}\,(f,b_\delta)b_\delta = -\gamma^2 f \ ,$$

where we have set $b_\delta = P_{F_\delta} b$. Thus

$$f = (f,b_\delta)\,\frac{1}{\pi}(X^2_{\mu_0} + \gamma^2)^{-1}\, b_\delta \ .$$

Taking inner products with b_δ, and using the obvious fact that (f,b_δ) cannot be zero, we obtain

$$1 = \frac{1}{\pi}\int_{F_\delta} \frac{|b(t)|^2}{(t-\mu_0)^2+\gamma^2}\, dt \ . \tag{5.9}$$

Condition (5.6) implies that (5.9) cannot hold for δ sufficiently small.

From now on we will assume that δ is a fixed positive number satisfying

$$\sup_{\gamma>0} \frac{1}{\pi}\int_{F_\delta} \frac{|b(t)^2|}{(t-\mu_0)^2+\gamma^2}\, dt < 1. \tag{5.10}$$

In particular, the operator R_δ defined in (5.8) is non-negative. From (5.7), for any $\varepsilon > 0$,

$$H_\lambda H_\lambda^* + \varepsilon I \geq R_\delta + \varepsilon I \geq \varepsilon I.$$

Applying Lemma 4.1, we obtain

$$((H_\lambda H_\lambda^* + \varepsilon)^{-1}\, b_\delta, b_\delta) \leq ((R_\delta+\varepsilon I)^{-1}\, b_\delta, b_\delta).$$

A direct computation of $(R_\delta+\varepsilon I)^{-1}\, b_\delta$ shows

$$((R_\delta+\varepsilon I)^{-1}\, b_\delta, b_\delta) = \frac{\displaystyle\int_{F_\delta} \frac{|b(t)|^2}{(t-\mu_0)^2+\varepsilon}\, dt}{\displaystyle 1-\frac{1}{\pi}\int_{F_\delta} \frac{|b(t)^2|}{(t-\mu_0)^2+\varepsilon}\, dt} \quad .$$

It follows from (5.10), that when $\lambda = \mu_0 + i\nu$,

$$\sup_{\varepsilon>0} ((H_\lambda H_\lambda^* + \varepsilon)^{-1}\, b_\delta, b_\delta) \leq M < \infty \quad .$$

We remark that the constant M does not depend on ν.

Proceeding as in the proof of Theorem 4.1, we obtain a weakly continuous $L^2(E)$-valued function $b_\delta(\lambda)$, defined for $\lambda = \mu_0 + i\nu$ $(-\infty < \nu < \infty)$, satisfying

$$(H-\lambda)b_\delta(\lambda) = b_\delta \ , \quad \|b_\delta(\lambda)\| \leq M \ .$$

Let Γ be a circle centered at μ_0 such that $\sigma(H)$ is interior to and let Γ_1 and Γ_2 be the two semicircles determined by Γ and the line $\mathrm{Re}\,\lambda = \mu_0$. Then

$$b_\delta = -\frac{1}{2\pi i}\int_\Gamma b_\delta(\lambda)d\lambda$$

$$= -\frac{1}{2\pi i}\int_{\Gamma_1} b_\delta(\lambda)d\lambda - \frac{1}{2\pi i}\int_{\Gamma_2} b_\delta(\lambda)d\lambda .$$

Consequently, one of these latter integrals is non-zero.

Suppose, for definiteness, $b_1 = -\frac{1}{2\pi i}\int_{\Gamma_1} b_\delta(\lambda)d\lambda \neq 0$. By a now familiar argument, we can conclude $\sigma_H(b_1)$ is contained in the closed semi-disc with boundary Γ_1. The hypothesis (5.5) implies $\sigma_H(b_1)$ is a proper subset of $\sigma(H)$. By Corollary 3.2, the operator H has a non-trivial invariant subspace. This completes the proof of the proposition.

A complete description of the spectrum of the operator H defined by equation (5.4) appears in Clancey and Putnam [1] and Pincus [2]. See also the remark following Proposition 4.1 in Chapter 5 of these notes. It is left to the discretion of the reader to make interesting choices of the functions a,b and the set E to produce non-trivial local spectral subspaces for the operator H as described in the proof of Proposition 5.2.

We now turn to the second applicaton of local spectral theory. As remarked earlier, the result derived below will be used in Chapter 3.

Let us agree to say that a subset L of a compact set K in the plane is contained in an exposed line segment in case there is an open disc Δ such that $\Delta \cap K = L$ and L is contained in a diameter of Δ.

Proposition 5.3. Let H be a pure hyponormal operator on $\mathcal{H}$. Then no non-empty subset of $\sigma(H)$ is contained in an exposed line segment.

Proof. Suppose the conclusion of the proposition is false. Without loss of generality, it can be assumed that H is a pure hyponormal operator whose spectrum intersects the open unit disc only along the y-axis 0_y and further, this intersection is non-empty. Let α,β,γ be three points in this intersection. (It is trivial to dispense with the case where the intersection does not contain at least three points.) It can be assumed that γ is on the line segment joining α and β. Let $g(z) = (z-\alpha)(z-\beta)$ and Γ be the circle centered on the axis 0_y passing through α,β.

The growth condition on the resolvent (Corollary 1.1) shows that, for every $f \in \mathcal{H}$, the function

$$\tilde{f}(z) = g(z)(H-z)^{-1}f, \quad z \in \Gamma\backslash\{\alpha,\beta\} \quad ,$$

extends to be weakly continuous on Γ. For $f \in \mathcal{H}$, define

$$\tilde{f} = \frac{1}{2\pi i}\int_\Gamma g(z)(H-z)^{-1}f\,dz \; ,$$

where the integral is a weak integral. It is easy to see, for every $f \in \mathcal{H}$, $\sigma_H(\tilde{f}) \subset 0_y$. Further, if $f \notin R(H-\gamma)$, then $\gamma \in \sigma_H(\tilde{f})$.

Thus the subspace $\mathcal{M} = \mathcal{M}_H(0_y)$ is a non-zero invariant subspace for H (see Theorem 3.2). The operator H restricted to $\mathcal{M}$ has spectrum in 0_y. It follows from Proposition 5.1 that $H|_{\mathcal{M}}$ is normal. This contradicts the purity of H and completes the proof.

Notes

Section 1. For another description of the elementary properties of seminormal operators the reader is referred to the book of Putnam [1]. Lemma 1.1 is a special case of a well known result of Douglas [1]. Lemma 1.2 is stated in Stampfli [8]. Proposition 1.2 and Corollary 1.1 are in Stampfli [1,2]. The result in Theorem 1.1 is due to Putnam [1].

Section 2. The study of subnormal operators is growing by leaps and bounds. The reader is referred to the development of the functional calculus for subnormal operators in Conway and Olin [1] and the most striking proof of the existence of invariant subspaces for subnormal operators by Brown [1]. Operators of the form (2.2) were first studied by Xa Dao-xeng [1]. The self-adjoint singular integral operators which appear as the imaginary parts of the operators (2.2) were diagonalized in Pincus [1] and also in Rosenblum [1].

Section 3. The local spectral theory for hyponormal operators has been developed by Stampfli [4-8]. The results in Lemmas 3.2, 3.3, 3.4 and Theorems 3.1, 3.2 are in Stampfli [3,8]. The original version of Stampfli's Theorem 3.2 required that the spectrum of H consist only of continuous spectrum. The improved version as stated in Theorem 3.2 appears in Radjabalipour [1]. Theorem 3.3 is in Putnam [6] and Johnson [1]. Theorem 3.4 was first proved in Radjabalipour [1]. Theorem 3.5 is in Clancey [2].

Section 4. Lemma 4.1 is a special case of a result of Loewner [1]. Theorem 4.1 and Corollary 4.2 are due to Putnam [8]. See also Radjabalipour [2]. Theorem 4.2 appears as a scholium in Stampfli [8].

Section 5. A complete proof of Proposition 5.1 is in Putnam [1]. A result which generalizes Proposition 5.2 appears in Apostol and Clancey [1]. Putnam [12] gives a result analogous to Proposition 5.2 for general seminormal operators. Stronger versions of Proposition 5.3 are derived in Stampfli [7].

CHAPTER 2

CONCRETE REALIZATIONS OF SEMINORMAL OPERATORS

In this chapter we follow an idea of Paul Muhly [1] to provide a singular integral representation for seminormal operators. Other authors, notably J. D. Pincus [1] and T. Kato [2], have earlier given independent derivations of this representation. Following Muhly's method all the way to its natural conclusion, we obtain several of the basic results concerning seminormal operators. As a consequence of this singular integral representation, we will deduce the self-adjoint commutator inequalities of C. R. Putnam and T. Kato.

1. Symbols and Γ-operators.

This section is concerned with the usual paraphernalia connected with self-adjoint commutator equations. The main ingredients are the Friedrichs Γ-operators and the symbol homomorphisms. Fortunately, for the case of a non-negative commutator, the existence of these objects is easily obtained. As a consequence, this section is concerned mostly with algebraic properties of the operations.

Let A be a fixed self-adjoint operator on $\mathcal{H}$. We define the symbols $S_A^{\pm}(T)$ of the operator T on $\mathcal{H}$ with respect to A as the operators

$$S_A^{\pm}(T) = \underset{t\to\pm\infty}{\text{s-lim}}\, e^{itA}\, T e^{-itA}, \qquad (1.1)$$

whenever these strong limits exist. It is important to note that there are two distinct symbols defined in (1.1). In fact, as we will see, many invariants of the operator T can be detected by the difference $S_A^+(T) - S_A^-(T)$.

The following proposition summarizes some of the algebraic properties of the symbols:

Proposition 1.1. Let A be a fixed self-adjoint operator on $\mathcal{H}$ and $S_A^{\pm}$ the operations defined by (1.1). Let T be in $L(\mathcal{H})$ and assume $S_A^{\pm}(T)$ exist. Then

(i) $$AS_A^{\pm}(T) = S_A^{\pm}(T)A, \text{ and}$$

(ii) $$\|S_A^{\pm}(T)\| \le \|T\| .$$

(iii) If $S_A^{\pm}(T^*)$ exist, then

$$[S_A^{\pm}(T)]^* = S_A^{\pm}(T^*) \quad .$$

Suppose T_1, T_2 are in $L(\mathcal{H})$ and $S_A^{\pm}(T_i)$ exist $(i = 1,2)$.

(iv) For $\alpha,\beta \in \mathbb{C}$, $S_A^{\pm}(\alpha T_1 + \beta T_2)$ exist and

$$S_A^{\pm}(\alpha T_1 + \beta T_2) = \alpha S_A^{\pm}(T_1) + \beta S_A^{\pm}(T_2)$$

(v) The symbols $S_A^{\pm}(T_1T_2)$ exist and

$$S_A^{\pm}(T_1T_2) = S_A^{\pm}(T_1)S_A^{\pm}(T_2).$$

Proof. The majority of the statements (i)-(v) are trivial. For (i) we note, if λ is a real number, then

$$e^{i\lambda A}[e^{itA}Te^{-itA}] = [e^{i(t+\lambda)A}Te^{-i(t+\lambda)A}]e^{i\lambda A}.$$

Consequently, $e^{i\lambda A}S_A^{\pm}(T) = S_A^{\pm}(T)e^{i\lambda A}$, $-\infty < \lambda < \infty$. Differentiating this last identity with respect to λ we obtain

$$iAe^{i\lambda A}S_A^{\pm}(T) = i\, S_A^{\pm}(T)Ae^{i\lambda A}.$$

The result in (i) now follows by setting $\lambda = 0$. The verifications of the remaining statements of the proposition are left to the reader.

Proposition 1.1 shows that the operations $S_A^{\pm}$ are contractive "star" homomorphisms into the commutant of the operator A.

Again let A be a fixed self-adjoint operator on $\mathscr{H}$. The Friedrichs Γ-operators are the operators $\Gamma_A^{\pm}$ defined for $T \in L(\mathscr{H})$ by

$$\Gamma_A^{\pm}(T) = \pm\int_0^{\pm\infty} e^{itA}Te^{-itA}dt$$

$$= \underset{s\to+\infty}{\text{s-lim}} \pm \int_0^{\pm s} e^{itA}Te^{-itA}dt \,, \qquad (1.2)$$

whenever these strong limits exist.

The important role played by the Γ-operations in the study of commutator equations is contained in the following:

Proposition 1.2. Let A be a fixed self-adjoint operator on $\mathscr{H}$. Assume $D \in L(\mathscr{H})$ is such that $\Gamma_A^{\pm}(D)$ exist. Then

$$i[A,\Gamma_A^{\pm}(D)] = i[A\Gamma_A^{\pm}(D) - \Gamma_A^{\pm}(D)A] = \mp D \,. \qquad (1.3)$$

Proof. Let λ be a real number. Then

$$e^{i\lambda A}\Gamma_A^{\pm}(D)e^{-i\lambda A} = \pm\int_0^{\pm\infty} e^{i(\lambda+t)A}De^{-i(\lambda+t)A}dt$$

$$= \pm\int_{\lambda}^{\pm\infty} e^{itA}De^{-itA}dt \quad .$$

Differentiating (in the strong sense) both sides of the last identity with respect to λ, we obtain

$$e^{i\lambda A}i[A\Gamma_A^{\pm}(D) - \Gamma_A^{\pm}(D)A]e^{-i\lambda A} = \mp e^{i\lambda A}De^{-i\lambda A} \quad .$$

Setting $\lambda = 0$ in this last equality, we obtain (1.3). This completes the proof of the proposition.

Next we relate the operations $S_A^{\pm}$ and $\Gamma_A^{\pm}$. Obviously,

$$\frac{d}{dt}\, e^{itA}Te^{-itA} = e^{itA}i[A,T]e^{-itA}. \qquad (1.4)$$

This leads to the following:

Proposition 1.3. Let A be a fixed self-adjoint operator and $T \in L(\mathcal{H})$. The symbols $S_A^{\pm}(T)$ exist if and only if $\Gamma_A^{\pm}(i[T,A])$ exist. Moreover, whenever one of $S_A^{\pm}(T)$ or $\Gamma_A^{\pm}(i[T,A])$ exist, then

$$T = S_A^{\pm}(T) \pm \Gamma_A^{\pm}(i[T,A]) \ . \tag{1.5}$$

Proof. From (1.4) we have, for s a real number,

$$e^{isA}Te^{-isA} - T = \int_0^s e^{itA} i[A,T]e^{-itA} dt \ .$$

The validity of the proposition clearly follows from this last identity. This completes the proof.

Corollary 1.1. Suppose A is self-adjoint, $D \in L(\mathcal{H})$, and $\Gamma_A^{\pm}(D)$ exist. Then both $S_A^{\pm}(D)$ and $S_A^{\pm}(\Gamma_A^{\pm}(D))$ exist and are zero.

Proof. The existence of $\Gamma_A^{\pm}(D)$ implies the existence of $\Gamma_A^{\pm}(i[A,D])$. By Proposition 1.2,

$$i[A,\Gamma_A^{\pm}(D)] = \Gamma_A^{\pm}(i[A,D]) = \mp D.$$

From Proposition 1.3, $S_A^{\pm}(D)$ exist and

$$D = S_A^{\pm}(D) \pm \Gamma_A^{\pm}(i[D,A])$$

$$= S_A^{\pm}(D) + D.$$

This shows $S_A^{\pm}(D) = 0$.

Further, if $\Gamma_A^{\pm}(D)$ exist, then $i[\Gamma_A^{\pm}(D),A] = \pm D$ (see equation (1.3)). Consequently, $\Gamma_A^{\pm}(i[\Gamma_A^{\pm}(D),A])$ exist and equal $\pm\Gamma_A^{\pm}(D)$. From Proposition 1.3, $S_A^{\pm}(\Gamma_A^{\pm}(D))$ exist and using (1.5),

$$\Gamma_A^{\pm}(D) = S_A^{\pm}(\Gamma_A^{\pm}(D)) + \Gamma_A^{\pm}(D) \ .$$

This implies $S_A^{\pm}(\Gamma_A^{\pm}(D)) = 0$ and completes the proof.

It is now easy to discuss self-adjoint solutions Z of the commutator equation

$$i[AZ - ZA] = R^2 , \tag{1.6}$$

where A, R are fixed self-adjoint operators on $\mathcal{H}$.

<u>Theorem 1.1</u>. Assume the commutator equation (1.6) has a solution Z_0. Then $S_A^{\pm}(Z_0)$ and $\Gamma_A^{\pm}(R^2)$ exist; moreover,

$$Z_0 = S_A^{\pm}(Z_0) \mp \Gamma_A^{\pm}(R^2). \tag{1.7}$$

The general solution of (1.6) is of the form

$$Z = B \mp \Gamma_A^{\pm}(R^2) ,$$

where B is an arbitrary operator commuting with A. The solutions $Q^{\pm} = \mp\Gamma_A^{\pm}(R^2)$ are the unique solutions of (1.6) satisfying $S_A^{\pm}(Q^{\pm}) = 0$.

Proof. Differentiating the function $e^{itA}Z_0e^{-itA}$, we obtain

$$\frac{d}{dt}[e^{itA}Z_0e^{-itA}] = e^{itA}i[AZ_0-Z_0A]e^{-itA} = e^{itA}R^2e^{-itA} \geq 0.$$

It follows that $e^{itA}Z_0e^{-itA}$ is an increasing bounded family of self-adjoint operators on the real line. Therefore, $S_A^{\pm}(Z_0)$ exist. Immediately, from Proposition 1.3, $\Gamma_A^{\pm}(R^2)$ exist and (1.7) holds. The form of the general solution of (1.6) as stated in the theorem is obvious. The last statement of the theorem follows from Corollary 1.1. This completes the proof.

An immediate corollary of Theorem 1.1 is the following:

Corollary 1.2. Let $H = X+iY$ be a hyponormal operator on $\mathcal{H}$ with self-commutator $[H^*,H] = 2R^2 \geq 0$. Then $S_X^{\pm}(Y)$ and $\Gamma_X^{\pm}(R^2)$ exist and

$$H = X + i[S_X^{\pm}(Y) \mp \Gamma_X^{\pm}(R^2)] \ . \tag{1.8}$$

In Section 3 we will obtain a more explicit representation for H when X is diagonalized.

Let T be an operator on $\mathcal{H}$. The notation $\mathcal{C}^*(T)$ will be employed for the $\mathcal{C}^*$-algebra generated by T and the identity. If $T = X+iY$ is the Cartesian form of T, then obviously $\mathcal{C}^*(T)$ is the $\mathcal{C}^*$-algebra (with identity) generated by the two self-adjoint operators X and Y.

Corollary 1.3. Let $H = X + iY$ be the Cartesian form of the hyponormal operator H on $\mathcal{H}$. The symbol maps $S_X^{\pm}$ can be extended to star homomorphisms of $\mathcal{C}^*(H)$ into the commutant of X.

Proof. In Corollary 1.2 it was observed that $S_X^{\pm}(Y)$ exist. Obviously, $S_X^{\pm}(X) = X$. It is clear from the properties of $S_X^{\pm}$ stated in Proposition 1.1 that $S_X^{\pm}$ extend to star-homomorphisms from $\mathcal{C}^*(H)$ into the commutant of X. This completes the proof.

Remarks. Let $H = X + iY$ be a hyponormal operator and $S_X^{\pm}$ the star homomorphisms of $\mathcal{C}^*(H)$ described in Corollary 1.3.

1°. The images of the hyponormal operator H under $S_X^{\pm}$ are the normal operators

$$N_{\pm} = X + iS_X^{\pm}(Y). \tag{1.9}$$

2°. It is well known that the spectrum of the operator H in $L(\mathcal{H})$ is the same as the spectrum of H considered as an element in $\mathcal{C}^*(H)$. Since $S_X^{\pm}$ are star homorphisms,

$$\sigma(N_{\pm}) \subset \sigma(H) \ . \tag{1.10}$$

3°. The inclusion (1.10) can be used to establish the inclusion

$$\sigma(X) \subset \mathrm{proj}_X \sigma(H) , \qquad (1.11)$$

which was left unproved in Proposition 5.1 of Chapter 1. In fact, it is well known that for the normal operators $N_\pm$ defined in (1.9)

$$\sigma(X) = \mathrm{proj}_X(\sigma(N_\pm)) . \qquad (1.12)$$

This result can easily be established for an arbitrary normal operator N by studying the relation between the spectral resolutions of N and its real part. The inclusion (1.11) is immediate from (1.10) and (1.12).

4°. The star homorphisms $S_X^\pm$ annihilate the commutator ideal $\mathcal{Q}$ in $\mathcal{C}^*(H)$ (this ideal coincides with the smallest two sided star ideal containing $[H^*,H]$). In general, the kernel of $S_X^\pm$ is larger than $\mathcal{Q}$. For some further results on the algebra $\mathcal{C}^*(H)/\mathcal{Q}$, the reader is referred to Howe [1].

2. Diagonalization of Self-Adjoint Operators.

In this section we briefly introduce notation so that a convenient version of "the spectral multiplicity theorem" can be formulated for self-adjoint operators. These notations will be used in the sequel. No attempt is made to prove anything. The details of the description can be found, for example, in Dixmier [1].

It is assumed that $\mathcal{H}$ is a separable Hilbert space. Let A be a self-adjoint operator on $\mathcal{H}$ with the spectral resolution $A = \int t dE(t)$. For any vector $f \in \mathcal{H}$, we introduce the Borel measure ν_f defined on the real line by $d\nu_f(t) = d\|E(t)f\|^2$. A <u>scalar spectral measure</u> for the operator A is a non-negative Borel measure ν on $\mathbb{R}$ such that ν_f is absolutely continuous with respect to ν, for every $f \in \mathcal{H}$, and if β is a Borel set with $\nu_f(\beta) = 0$, for every $f \in \mathcal{H}$, then $\nu(\beta) = 0$. If $\{\varphi_j\}_{j=1}^\infty$ is a complete orthonormal system in $\mathcal{H}$, then $\nu = \sum_{j=1}^\infty 2^{-j}\nu_{\varphi_j}$ is a scalar spectral measure for A. It is easy to see that any two scalar spectral measures are mutually absolutely continuous.

The absolutely continuous subspace of the operator A is the collection $\mathcal{H}_{ac}(A)$ of vectors $f \in \mathcal{H}$, such that ν_f is absolutely continuous with respect to Lebesgue measure on $\mathbb{R}$. It is well known (and easy to prove) that $\mathcal{H}_{ac}(A)$ is a closed A invariant subspace (see e.g. Kato [1]) The orthocomplement $\mathcal{H}_s(A) = [\mathcal{H}_{ac}(A)]^{\perp}$ is referred to as the singular subspace of A. Corresponding to the decomposition $\mathcal{H} = \mathcal{H}_{ac}(A) \oplus \mathcal{H}_s(A)$, we have the decomposition of the operator $A = A_{ac} \oplus A_s$ into its absolutely continuous and singular components.

It should be noted that in the definitions (1.1) and (1.2) of the symbol and Γ-operations, one usually restricts the operators to $\mathcal{H}_{ac}(A)$. Later in this chapter, it will be shown that if $T = X + iY$ is a pure semi-normal operator, then $\mathcal{H}_{ac}(X) = \mathcal{H}_{ac}(Y) = \mathcal{H}$. This result partially explains why there is no need to restrict the definitions in (1.1) and (1.2) to $\mathcal{H}_{ac}(X)$.

The description of the diagonalization of self-adjoint operators will involve direct integral notation. Let $\mathcal{H}_\infty$ denote a fixed separable Hilbert space and let $\{\mathcal{H}_j\}_{j=0}^{\infty}$ be a chain of subspaces of $\mathcal{H}_\infty$ such that dimension $\mathcal{H}_j$ equals j and $\bigcup_{j=0}^{\infty} \mathcal{H}_j$ spans $\mathcal{H}_\infty$. We have

$$(0) = \mathcal{H}_0 \subset \mathcal{H}_1 \subset \mathcal{H}_2 \subset \ldots \subset \mathcal{H}_\infty. \tag{2.1}$$

For every real t, let $\mathcal{H}(t)$ be one of the subspaces in (2.1) having dimension n(t). It will be assumed that

$$E_j = \{t : \dim \mathcal{H}(t) = j\} , \quad 1 \le j \le \infty \quad ,$$

are Borel sets. The notation $F = \bigcup_{j>0} E_j$ will be employed.

Let ν be a Borel measure on the real line $\mathbb{R}$. By the direct integral space

$$\tilde{\mathcal{H}} = \int_{\mathbb{R}} \oplus \mathcal{H}(t) d\nu \ , \tag{2.2}$$

we will mean the equivalence classes of functions f on $\mathbb{R}$ such that:

(i) f(t) is a weakly ν-measurable function with values in $\mathcal{H}_\infty$.

(ii) If t is in E_j, then f(t) is in $\mathcal{H}_j$ (ν a.e.).

(iii) $\int \|f(t)\|^2 d\nu(t) < \infty$, where $\| \ \|$ is the norm in $\mathcal{H}_\infty$.

The space $\tilde{\mathcal{H}}$ is a Hilbert space furnished with the inner product

$$(f,g) = \int_{\mathbb{R}} (f(t), g(t))\, d\nu(t) .$$

The Hilbert space $\tilde{\mathcal{H}}$ is called a direct integral space with measure ν and <u>multiplicity function</u> $n = n(t)$.

There is a natural self-adjoint (possibly unbounded) operator Λ acting on the direct integral space (2.2). This is the operator defined on $\tilde{\mathcal{H}}$ by

$$\Lambda f(t) = t f(t) . \qquad (2.3)$$

A ν-measurable subset E will be called a <u>support set</u> for the direct integral space (2.2) in case

$$\nu(E \Delta F) = \nu(E \backslash F) \cup \nu(F \backslash E) = 0 . \qquad (2.4)$$

Obviously, (2.4) defines an equivalence class of ν-measurable sets. The operator Λ defined in (2.3) is bounded if and only if there is a bounded support set for the direct integral (2.2). If E is a support set for the direct integral space, then we will write

$$\tilde{\mathcal{H}} = \int_E \oplus \mathcal{H}(t)\, d\nu .$$

The spectral multiplicity theorem for self-adjoint operators will be formulated as follows:

<u>Theorem 2.1</u>. Let A be a self-adjoint operator on a Hilbert space $\mathcal{H}$. Then there is a unitary mapping U carrying $\mathcal{H}$ onto a direct integral space $\tilde{\mathcal{H}} = \int_E \oplus \mathcal{H}(t)\, d\nu$ such that $UAU^* = \Lambda$, where Λ is the operator defined on $\tilde{\mathcal{H}}$ by (2.3).

In the above theorem the measure ν is necessarily a scalar spectral measure for the operator A. In particular, if the operator A is absolutely continuous, then ν can be chosen as Lebesgue measure. The function $n(t) = \dim \mathcal{K}(t)$ (determined ν a.e.) is called the spectral multiplicity function of the operator A.

An operator B on the direct integral space $\tilde{\mathcal{K}} = \int_E \oplus \mathcal{K}(t)d\nu$ is called decomposable in case there are operators $B(t) \in L(\mathcal{K}(t))$ such that (i) For $f \in \tilde{\mathcal{K}}$, the function $B(t)f(t)$ is a weakly ν-measurable $\mathcal{K}_\infty$-valued function, (ii) $\text{esssup}\|B(t)\| < \infty$, here $\|B(t)\|$ refers to the norm of $B(t) \in L(\mathcal{K}(t))$, (iii) For $f \in \tilde{\mathcal{K}}$, $Bf(t) = B(t)f(t)$. Following the usual convention, the decomposable operator B is expressed in the form

$$B = \int_E \oplus B(t)d\nu . \tag{2.5}$$

A decomposable operator B obviously commutes with Λ. The converse of this last statement is also true. Any operator commuting with Λ is decomposable. The norm of the decomposable operator B of (2.5) satisfies

$$\|B\| = \text{esssup}\|B(t)\| .$$

3. Singular Integral Representations of Seminormal Operators.

In this section we will provide a singular integral representation for seminormal operators acting on a separable Hilbert space. This representation occurs when the real part is diagonalized on a direct integral Hilbert space. The argument is basically due to Muhly [1]; however, in the description below we include the contributions made by the singular component of the diagonalization.

The notation , conventions and results of this section will be used freely in the remainder of these notes. The representation of the seminormal operator proceeds in a sequence of steps which we finally summarize in Theorem 3.1. At the end of this section we deduce a result of Putnam [2] which establishes that the real and imaginary parts of a pure seminormal operator are absolutely continuous.

For definiteness, it will be assumed that $H = X+iY$ is a hyponormal operator on the Hilbert space $\mathcal{H}$. The operator X will be diagonalized. Trivial modifications of the discussion will handle the diagonalization of the operator Y (e.g., consider the hyponormal operator $H' = iH$). At this stage we do not assume the operator H is pure.

The commutator $i[X,Y]$ will be assumed of the form

$$i[XY-YX] = R^2 , \tag{3.1}$$

where R is a non-negative operator. The closure of the range of R^2 will be denoted by $\mathcal{R}$. The notation $\mathcal{H}_1$ will be employed for the smallest subspace of $\mathcal{H}$ reducing X and containing the space $\mathcal{R}$. Specifically,

$$\mathcal{H}_1 = \bigvee\{\mathcal{R}, X\mathcal{R}, X^2\mathcal{R}, \ldots\} \quad . \tag{3.2}$$

The notation X_1 will be used for the restriction of the operator X to the reducing subspace $\mathcal{H}_1$.

Step 1. The diagonalization of X_1. Let $L^2(\mathbb{R}:\mathcal{R})$ denote the Hilbert space of weakly Lebesgue measurable $\mathcal{R}$-valued square integrable functions on $\mathbb{R}$. The space $L^2(\mathbb{R}:\mathcal{R})$ can be viewed as the direct integral space $\int_{\mathbb{R}} \oplus \mathcal{H}(t)dt$, where $\mathcal{H}(t) = \mathcal{R}$. Accordingly, we can consider the (unbounded) self-adjoint operator Λ defined by (2.3) on $L^2(\mathbb{R}:\mathcal{R})$.

Define the operator $J:\mathcal{H} \to L^2(\mathbb{R}:\mathcal{R})$ by

$$Jf(t) = Re^{-itX}f, \; t \in \mathbb{R} . \tag{3.3}$$

In Section 1 of this chapter (see Theorem 1.1) it was observed

$$\Gamma_X(R^2) \equiv \Gamma_X^+(R^2) + \Gamma_X^-(R^2) = \int_{-\infty}^{\infty} e^{itX}R^2e^{-itX}dt ,$$

considered as a strong operator integral, defines a bounded operator on $\mathcal{H}$. For f in $\mathcal{H}$,

$$\|Jf\|^2 = \int_{-\infty}^{\infty} \|Re^{-itX}f\|^2 dt = (\Gamma_X(R^2)f,f) .$$

Consequently, the operator J defined in (3.3) is a bounded operator.

The space $\mathcal{K}_1$ coincides with the initial space of the operator J. We verify this last statement as follows. A vector $f \in \mathcal{K}$ is in the kernel of J if and only if for all $t \in \mathbb{R}$ and $g \in R$

$$(Re^{-itX}f,g) = (f,e^{itX}Rg) = 0.$$

This is obviously equivalent to the condition that f belongs to $\mathcal{K} \ominus \mathcal{K}_1$. We will use the notation

$$\mathcal{K}_2 = \mathcal{K}_1^{\perp} \quad . \tag{3.4}$$

For any real number s, we define the unitary operator $U_s = e^{isX}$ on $\mathcal{K}$ and we will let T_s denote the translation operator defined on $L^2(\mathbb{R}:\mathcal{R})$ by

$$T_s f(t) = f(t-s) .$$

Then $JU_s f(t) = Re^{-itX}e^{isX}f = Re^{-i(t-s)X}f = T_s Jf(t)$. This last computation establishes the intertwining identity

$$JU_s = T_s J, \quad s \in \mathbb{R} . \tag{3.5}$$

Note that (3.5) implies the closure of the range of J is invariant under the unitary group $\{T_s\}_{s\in\mathbb{R}}$. The closure of the range of J will be denoted by $\mathcal{L}_1$. The subspace $\mathcal{L}_1$ is also the closure of the range of J restricted to its initial space $\mathcal{K}_1$.

Let J = WK be the polar factorization of the operator J. Then $K = (J^*J)^{1/2}$, where $J^*:L^2(\mathbb{R}:\mathcal{R}) \to \mathcal{K}$ is the adjoint of J and W is a partial isometry with initial space $\mathcal{K}_1$ and final space $\mathcal{L}_1$.

From (3.5), we know that K commutes with $\{U_s\}_{s\in\mathbb{R}}$. Taking advantage of the fact that ker K = kerJ = $\mathcal{K}_2$, we can write

$$W_1(U_s)_1 = (T_s)_1 W_1 , \quad s \in \mathbb{R} , \tag{3.6}$$

where $(U_s)_1 = U_s|_{\mathcal{K}_1}$, $(T_s)_1 = T_s|_{\mathcal{L}_1}$ and W_1 is the unitary operator W carrying $\mathcal{K}_1$ onto $\mathcal{L}_1$.

Actually, at this stage we could appeal to the absolute continuity of the unitary group $\{T_s\}_{s\in\mathbb{R}}$ to conclude (from (3.6)) the absolute continuity of X_1. Our purposes will be better served if we first take the Fourier transform.

Digression. The vector valued Fourier transform. The method used to define the Fourier transform on $L^2(\mathbb{R}:\mathcal{K})$ is the usual method which is used to lift an operator initially defined on $L^2(\mathbb{R})$ to its vector valued counterpart on $L^2(\mathbb{R}:\mathcal{K})$. We will briefly describe this abstract procedure. We emphasize that we are not concerned with vector questions concerning the a.e. existence of an operator like the Fourier transform on $L^2(\mathbb{R}:\mathcal{K})$. Rather, we are interested only in the fact that the lift preserves the algebraic properties possessed by the operator on $L^2(\mathbb{R})$. The process we are describing is simply the process of tensoring with the identity operator of $\mathcal{K}$.

Let T be a bounded operator on $L^2(\mathbb{R})$. Let $\{\varphi_i\}_{i=1}^{\infty}$ be a complete orthonormal basis in $L^2(\mathbb{R})$ and $\{\Psi_j\}_{j=1}^{N}$ (N=dimension $\mathcal{K}$) be a complete orthonormal system in $\mathcal{K}$. The functions $f_{ij} = \varphi_i\Psi_j (\in L^2(\mathbb{R}:\mathcal{K}))$ $i=1,2,\ldots$; $j=1,2,\ldots,N$ form a complete orthonormal system in $L^2(\mathbb{R}:\mathcal{K})$. The lift of the operator T to $L^2(\mathbb{R}:\mathcal{K})$ is the operator $\hat{T}$ defined on $g = \sum_{i,j} g_{ij}\varphi_i\Psi_j$ by

$$\hat{T}g = \sum_{i,j} g_{ij}(T\varphi_i)\Psi_j .$$

The operator $\hat{T}$ is a bounded operator on $L^2(\mathbb{R}:\mathcal{K})$ with $\|\hat{T}\| = \|T\|$.

In particular, one can lift the Fourier transform F defined on $f \in L^2(\mathbb{R})$ by

$$F f(x) = \underset{b\to\infty}{\mathrm{l.i.m}} \frac{1}{\sqrt{2\pi}} \int_{-b}^{b} e^{-ixt} f(t)\,dt$$

to the vector valued Fourier transform defined on $L^2(\mathbb{R}:\mathcal{R})$. In this definition "l.i.m." denotes "limit in the mean". For a discussion of the classical Fourier transform the reader is referred to Titchmarsh [1] (see, also Rudin [1]). We will abuse the notation and write $\mathcal{F}$ (rather than $\hat{\mathcal{F}}$) for the vector valued Fourier transform on $L^2(\mathbb{R}:\mathcal{R})$.

The vector valued Fourier transform is a unitary mapping on $L^2(\mathbb{R}:\mathcal{R})$. As in the scalar case, we have

$$\mathcal{F}^*T_s = M_s\mathcal{F}^* \quad , \ s \in \mathbb{R}, \tag{3.7}$$

where M_s $(s \in \mathbb{R})$ is the decomposable operator defined on $L^2(\mathbb{R}:\mathcal{R})$ by

$$M_s f(t) = e^{ist} f(t).$$

This concludes our digression.

The partially isometric operator $V = \mathcal{F}^*W$ carries $\mathcal{K}_1$ isometrically onto a subspace $\mathcal{N}_1$ of $L^2(\mathbb{R}:\mathcal{R})$. Combining (3.6) and (3.7), we discover that this subspace is reducing for the group $\{M_s\}_{s\in\mathbb{R}}$. Moreover

$$V_1(U_s)_1 = (M_s)_1 V_1 , \tag{3.8}$$

where $(M_s)_1 = M_s|_{\mathcal{N}_1}$ and V_1 denotes the unitary operator $V: \mathcal{K}_1 \to \mathcal{N}_1$.

Let f be in $\mathcal{K}_1$. The vector function $V_1(U_s)_1 f$ possesses the strong derivative iV_1X_1f at $s = 0$. Directly from (3.8), it follows that $(M_s)_1V_1f$ has a strong derivative at $s = 0$. This derivative must be $i\Lambda V_1 f$. Indeed, $V_1X_1f = \Lambda V_1 f$, and this identity shows that the range of V_1 (which is $\mathcal{N}_1$) is a reducing subspace for Λ on which Λ defines a bounded operator. It develops that for some bounded Borel set E

$$\mathcal{N}_1 = \int_E \oplus\, \mathcal{N}_1(t)\,dt, \tag{3.9}$$

where $\mathcal{N}_1(t)$, for almost all $t \in E$, denotes a non-zero subspace of $\mathcal{R}$.

Set $\Lambda_1 = \Lambda|_{\mathcal{N}_1}$, then

$$V_1 X_1 = \Lambda_1 V_1 . \tag{3.10}$$

Equation (3.10) provides the diagonalization of the operator X_1 on the direct integral space (3.9).

Step 2. The representation of $\Gamma_X^{\pm}(R^2)$. In Corollary 1.2 it was established that the operator Y has the forms:

$$Y = S_X^{\pm}(Y) \mp \Gamma_X^{\pm}(R^2)$$

We are now in a position to discuss the representation of $\Gamma_X^{\pm}(R^2)$, when X is diagonalized.

The operators $\Gamma_X^{\pm}(R^2)$ are zero on the subspace $\mathcal{K}_2$. In fact, if g is in $\mathcal{K}_2$, then

$$(\Gamma_X^{\pm}(R^2)g,g) = \pm\int_0^{\pm\infty} (e^{itX}R^2 e^{-itX}g,g)\,dt = 0 .$$

Since, $\Gamma_X^{\pm}(R^2) \geq 0$, then $\Gamma_X^{\pm}(R^2)g = 0$. It follows that $\mathcal{K}_1$ reduces $\Gamma_X^{\pm}(R^2)$ and we will denote the restrictions of $\Gamma_X^{\pm}(R^2)$ to $\mathcal{K}_1$ by $[\Gamma_X^{\pm}(R^2)]_1$.

The adjoint of the operator J is formally given by

$$J^*f = \int_{-\infty}^{\infty} e^{itX}Rf(t)\,dt, \quad f \in L^2(\mathbb{R}:\mathcal{R}) . \tag{3.11}$$

In case f belongs to the range of J, the integral in (3.11) converges in the strong sense. In fact, $J^*J = \Gamma_X^{+}(R^2) + \Gamma_X^{-}(R^2)$. It is obvious from (3.11) and the definitions of $\Gamma_X^{\pm}(R^2)$ that

$$\Gamma_X^{\pm}(R^2) = J^*\chi_{\pm}J, \tag{3.12}$$

where χ_+ (respectively, χ_-) denotes the projection operator on $L^2(\mathbb{R}:\mathcal{R})$ of multiplication by the characteristic function of $\mathbb{R}^+ = (0,\infty)$ (respectively, $\mathbb{R}^- = (-\infty,0)$).

From (3.12) it follows that

$$\begin{aligned} V_1[\Gamma_X^{\pm}(R^2)]_1V_1^* &= V_1K_1V_1^*(F^*\chi_{\pm}F)|_{\eta_1}V_1K_1V_1^* \\ &= B_1P_{\pm}|_{\eta_1}B_1 , \end{aligned} \tag{3.13}$$

where we have set $B_1 = V_1K_1V_1^*$ and $P_\pm = F^*\chi_\pm F$. Note that in the last equation K_1 is just the operator K restricted to the reducing subspace $\mathcal{K}_1$, whereas, the notations $P_\pm|_{\mathcal{N}_1}$ denote the compressions of the projections $P_\pm$ to the subspace $\mathcal{N}_1$.

The operator B_1 commutes with Λ_1. This follows from equation (3.5). (We have earlier observed that K commutes with the group $\{U_s\}_{s\in\mathbb{R}}$.) Therefore, B_1 is decomposable and may be written in the form

$$B_1 = \int_E \oplus B_1(t)dt, \tag{3.14}$$

with $B_1(t) \geq 0$ almost everywhere.

Let σ denote the operator of multiplication by the function

$$\sigma(t) = \begin{cases} 1 & t \geq 0 \\ -1 & t < 0 \end{cases}$$

acting on $L^2(\mathbb{R})$. Then $F^*\sigma F = Q$, where Q denotes the Hilbert transform

$$Qf(x) = \frac{1}{\pi i}\int_{\mathbb{R}} \frac{f(t)}{t-x}\,dt, \quad f \in L^2(\mathbb{R}) .$$

This result is in Titchmarsh [1].

As a consequence, the operators $P_\pm = F^*\chi_\pm F$ on $L^2(\mathbb{R})$ have the form

$$P_\pm f(x) = \frac{1}{2}[f(x) \pm \frac{1}{\pi i}\int_{\mathbb{R}} \frac{f(t)}{t-x}\,dt], \quad f \in L^2(\mathbb{R}) . \tag{3.15}$$

As described in the above digression, we can lift the operators $P_\pm$ to their counterparts

$$\hat{P}_\pm = \frac{1}{2}[I \pm \hat{Q}]$$

on $L^2(\mathbb{R}:\mathcal{K})$. In the sequel, we will often write the operators $\hat{P}_\pm$ on $L^2(\mathbb{R}:\mathcal{K})$ in the singular integral form (3.15); however, this notation is merely a matter of convenience. At no stage do we interpret the operators $\hat{P}_{\pm}$ as vector valued singular integral operators.

The orthogonal projection P_1 of $L^2(\mathbb{R}:\mathcal{R})$ onto the subspace $\mathcal{H}_1$ has the decomposable form

$$P_1 = \int_E \oplus P_1(t)dt, \qquad (3.16)$$

where for almost all $t\in E$, $P_1(t)$ is an orthogonal projection onto $\mathcal{H}_1(t)$.

Putting together the pieces, we learn that the operators

$$\Gamma_1^{\pm} = V_1[\Gamma_X^{\pm}(R^2)]_1 V_1^* = \tfrac{1}{2}[B_1^2 \pm B_1P_1\hat{Q}P_1B_1]$$

have the singular integral representations

$$\Gamma_1^{\pm}f(x) = \tfrac{1}{2}[B_1^2(x)f(x) \pm \frac{B_1(x)P_1(x)}{\pi i}\int_E \frac{B_1(t)f(t)}{t-x}\,dt] \quad (3.17)$$

Step 3. The representation of $S_X^{\pm}(Y)$. The restriction of the operator X to the reducing subspace $\mathcal{K}_2$ will be denoted by X_2. It will be assumed that the operator X_2 is diagonalized on the direct integral space

$$\mathcal{H}_2 = \int \oplus \mathcal{H}_2(t)dv\ . \qquad (3.18)$$

We will let $V_2: \mathcal{K}_2 \to \mathcal{H}_2$ be the unitary operator implementing the diagonalization of X_2, so that,

$$V_2X_2 = \Lambda_2V_2\ ,$$

where Λ_2 denotes the operator Λ on the direct integral space (3.18).

The operator V_0 defined by

$$V_0 = V_1 \oplus V_2$$

is a unitary operator from $\mathcal{K} = \mathcal{K}_1 \oplus \mathcal{K}_2$ to the space

$$\mathcal{H}_0 = \mathcal{H}_1 \oplus \mathcal{H}_2\ .$$

The operator $\Lambda_0 = V_0 X V_0^*$ on $\mathcal{H}_0$ has the 2×2 matrix form

$$\Lambda_0 = \begin{bmatrix} \Lambda_1 & 0 \\ 0 & \Lambda_2 \end{bmatrix} . \tag{3.19}$$

Similarly the operators $\Gamma_0^{\pm} = V_0 \Gamma_X^{\pm}(R^2) V_0^*$ on $\mathcal{H}_0$ have the form

$$\Gamma_0^{\pm} = \begin{bmatrix} \Gamma_1^{\pm} & 0 \\ 0 & 0 \end{bmatrix} . \tag{3.20}$$

The operators $A_0^{\pm} = V_0 S_X^{\pm}(Y) V_0^*$ commute with Λ_0; hence, they have the form

$$A_0^{\pm} = \begin{bmatrix} A_1^{\pm} & A_{12}^{\pm} \\ A_{21}^{\pm} & A_2^{\pm} \end{bmatrix} ,$$

where $A_1^{\pm}$, $A_2^{\pm}$ are decomposable operators on $\mathcal{H}_1$, $\mathcal{H}_2$, respectively, and $A_{12}^{\pm}\Lambda_2 = \Lambda_1 A_{21}^{\pm}$, $A_{21}^{\pm}\Lambda_1 = \Lambda_2 A_{21}^{\pm}$.

From the fact that both the operators $S_X^{\pm}(Y) \mp \Gamma_X^{\pm}(R^2)$ equal Y (see, Corollary 1.2), and the form (3.20), we conclude $A_{12}^{+} = A_{12}^{-}$, $A_{21}^{+} = A_{21}^{-}$ and $A_2^{+} = A_2^{-}$. Consequently, the operators $A_0^{\pm}$ have the matrix representations

$$A_0^{\pm} = \begin{bmatrix} A_1^{\pm} & A_{12} \\ A_{12}^{*} & A_2 \end{bmatrix} , \tag{3.21}$$

where $A_1^{\pm}$, A_2 are decomposable operators on $\mathcal{H}_1$, $\mathcal{H}_2$, respectively, and $A_{12}:\mathcal{H}_2 \to \mathcal{H}_1$ satisfies $\Lambda_1 A_{12} = A_{12}\Lambda_2$.

The three steps outlined above are summarized in the following:

<u>Theorem 3.1</u>. Let $H = X + iY$ be a hyponormal operator on the separable space $\mathcal{K}$ with $i[XY-YX] = R^2$. Let $\mathcal{K}_1$ and $\mathcal{K}_2$ be the subspaces defined by (3.2) and (3.4), respectively. Let X_1 and X_2 denote the restrictions of X to the subspaces $\mathcal{K}_1$ and $\mathcal{K}_2$, respectively. The operator X_1 is diagonalized by V_1 as the absolutely continuous operator

on the direct integral space $\mathcal{H}_1$ of (3.9). Assume V_2 denotes the unitary operator implementing the diagonalization of X_2 on the direct integral space $\mathcal{H}_2$ of (3.18). The operator $V_0 = V_1 \oplus V_2$ is unitary from $\mathcal{K} = \mathcal{K}_1 \oplus \mathcal{K}_2$ to $\mathcal{H}_0 = \mathcal{H}_1 \oplus \mathcal{H}_2$ such that the operator $\tilde{H} = V_0 H V_0^*$ has the singular integral form

$$\tilde{H} = \Lambda_0 + i[A_0^{\pm} \mp \Gamma_0^{\pm}], \tag{3.22}$$

where, Λ_0, $A_0^{\pm}$, $\Gamma_0^{\pm}$ are defined in equations (3.19) - (3.21) .

The following corollaries of the representation (3.22) will be important in the sequel:

Corollary 3.1. Assume the notations of Theorem 3.1. The operator $R_0^2 = V_0 R^2 V_0^*$ has the form

$$R_0^2 = \begin{bmatrix} Z & 0 \\ 0 & 0 \end{bmatrix},$$

where Z is the integral operator defined on $\mathcal{H}_1$ by

$$Zf(x) = \frac{B_1(x)P_1(x)}{2\pi} \int_E B_1(t)f(t)dt. \tag{3.23}$$

Proof. The operator $R_0^2 = \frac{1}{2}[\tilde{H}^*, \tilde{H}]$. The computation of R_0^2 can be carried out with either choice of sign in the representation (3.22). For definiteness, we assume that $\tilde{H} = \Lambda_0 + i[A_0^+ - \Gamma_0^+]$. Then R_0^2 equals

$$i[\Lambda_0(A_0^+ - \Gamma_0^+) - (A_0^+ - \Gamma_0^+)\Lambda_0] = i[\Gamma_0^+\Lambda_0 - \Lambda_0\Gamma_0^+]$$

$$= \begin{bmatrix} i[\Gamma_1^+\Lambda_1 - \Lambda_1\Gamma_1^+] & 0 \\ 0 & 0 \end{bmatrix}$$

The result follows from the form (3.17) of Γ_1^+. This completes the proof.

The next corollary takes advantage of the fact that $S_X^{\pm}$ define two separate symbols:

Corollary 3.2. Assume the notations of Theorem 3.1. Then

$$A_0^+ - A_0^- = \begin{bmatrix} \int_E^{\oplus} B_1^2(t)\,dt & 0 \\ 0 & 0 \end{bmatrix} . \tag{3.24}$$

Proof. From Corollary 1.1, we have

$$S_{\Lambda_0}^-(\Gamma_0^-) = 0 .$$

Therefore,

$$S_{\Lambda_1}^-(B_1^2) = S_{\Lambda_1}^-(B_1 P_1 \hat{Q} P_1 B_1) = B_1^2 .$$

As a consequence,

$$\begin{aligned} A_0^- = S_{\Lambda_0}^-(A_0^- + \Gamma_0^-) &= S_{\Lambda_0}^-(A_0^+ - \Gamma_0^+) \\ &= A_0^+ - \begin{bmatrix} B_1^2 & 0 \\ 0 & 0 \end{bmatrix} . \end{aligned}$$

This completes the proof.

The representation of Theorem 3.1 is slightly unsatisfactory for the following reason. The separate diagonalizations of X_1 and X_2 do not give the diagonalization of X exactly as described in the spectral multiplicity theorem. We would have to rearrange $\mathcal{N} = \mathcal{N}_1 \oplus \mathcal{N}_2$ to give this diagonalization. For our purposes, the representation (3.22) is adequate.

If $\mathcal{H}_1 = \mathcal{H}$ and R^2 is one-dimensional, then the representation (3.22) (with minor adjustments) reduces to the example described in 4° of Section 2, Chapter 1.

We conclude this section with the following result of Putnam [2]:

Theorem 3.2. Let $S = X + iY$ be a pure seminormal operator. Then both X and Y are absolutely continuous self-adjoint operators.

Proof. As remarked earlier, it is sufficient to consider the case of a hyponormal operator $H = X + iY$ and to establish the absolute continuity of the operator X.

It can be assumed that $H = X + iY$ is represented as the operator $\tilde{H}$ in Theorem 3.1. Clearly, $\mathcal{H}_s(\Lambda_0) \subset \mathcal{H}_2$. Since $\Lambda_1 A_{12} = A_{12}\Lambda_2$, then $E_1(\beta)A_{12} = A_{12}E_2(\beta)$, for any Borel set β, where E_i denotes the spectral resolution of $\Lambda_i (i=1,2)$. We have noted above that the resolution E_1 is absolutely continuous. Thus $A_{12}\mathcal{H}_s(\Lambda_0) = \{0\}$ and, consequently, $\mathcal{H}_s(\Lambda_0)$ reduces $\tilde{H}$. This translates to the statement that $\mathcal{H}_s(X)$ reduces H and is orthogonal to the range of $[H^*,H]$. From Theorem 1.1 of Chapter 1, we conclude $\mathcal{H}_s(X) = \{0\}$. This completes the proof.

4. Basic Self-Adjoint Commutator Inequalities.

In this section we will apply the representation of Theorem 3.1 to obtain two separate self-adjoint commutator inequalities of Putnam [2] and Kato [2]. It is interesting that the first result derived below is due to Putnam [2] and was refined by Kato [2], whereas, the second inequality of Kato [2] was refined by Putnam [11].

Theorem 4.1. Let X,Y be a pair of self-adjoint operators on $\mathcal{H}$ with $i[XY - YX] = R^2 \geq 0$, where R is some non-negative operator. Then

$$\pi\|R^2\| \leq \text{meas}_1[\sigma(X)]\|Y\| , \qquad (4.1)$$

where meas_1 denotes one-dimensional Lebesgue measure.

Proof. Let $H = X + iY$. Then H is hyponormal and can be assumed to have the form described in (3.22) of Theorem 3.1.

From Corollary 3.1, we learn that $\|R^2\|$ equals the norm of the operator Z defined on $\mathfrak{N}_1$ by (3.23). Thus

$$\pi \|R^2\| \leq \text{meas}_1(E) \tfrac{1}{2} \{\operatorname*{esssup}_{t \in E} \|B_1(t)\|^2\} .$$

From Corollary 3.2,

$$\begin{aligned} \tfrac{1}{2} \text{ esssup} \|B_1(t)\|^2 &\leq \tfrac{1}{2}\|A_0^-\| + \tfrac{1}{2}\|A_0^+\| \\ &= \tfrac{1}{2}\|S_X^+(Y)\| + \tfrac{1}{2}\|S_X^-(Y)\| \\ &\leq \|Y\| . \end{aligned}$$

Consequently,

$$\pi \|R^2\| \leq \text{meas}_1(E)\|Y\| , \tag{4.2}$$

which clearly implies (4.1). This completes the proof.

The reader will note that (4.2) is sharper than (4.1). The fact that (4.1) can be improved to (4.2) was noted by Kato [2].

Let A be a self-adjoint operator on $\mathcal{H}$ with scalar spectral measure ν and multiplicity function n. The multiplicity function n is called summable in case

$$\int n \, d\nu < \infty .$$

Let $H = X + iY$ be hyponormal with $i[XY - YX] = R^2$. Let $\mathcal{H}_1$ be the space defined in (3.2). Let n (respectively, n_1) denote the spectral multiplicity function of X (respectively, X_1). The following result was first established by Kato [2] using the multiplicity function n. The improvement to the present result is in Putnam [11].

Theorem 4.2. Let $H = X + iY$ be a hyponormal operator with $i[XY - YX] = R^2$. Let n_1 denote the spectral multiplicity function of the operator X restricted to the space $\mathcal{H}_1$ of (3.2). The following estimate holds

$$\pi \operatorname{tr}[R^2] \leq \|Y\| \int_{\mathbb{R}} n_1(t)dt , \qquad (4.3)$$

where $\operatorname{tr}[R^2]$ denotes the trace of R^2. In particular, if the spectral multiplicity function n_1 is summable, then the self-commutator of H is trace class.

The proof of Theorem 4.2 requires a lemma. The reader who requires more information concerning the trace can consult Gohberg and Krein [1] and Schatten [1]. (See, also Section 1 of Chapter 4.)

Lemma 4.1. Let $\tilde{\mathcal{H}} = \int_{\mathbb{R}} \oplus \mathcal{H}(t)d\nu$ denote a direct integral space with $\int d\nu < \infty$. Assume $\tilde{\mathcal{H}}$ is presented as the range of the decomposable projection $\int \oplus P(t)d\nu$ acting on $L^2(\mathbb{R}:\mathcal{H}_\infty:d\nu)$. Let $\int_{\mathbb{R}} \oplus K(t)d\nu$ be a decomposable operator on $\tilde{\mathcal{H}}$. If K denotes the integral operator defined on $\tilde{\mathcal{H}}$ by

$$Kf(x) = K^*(x)P(x)\int_{\mathbb{R}} K(t)f(t)d\nu \quad ,$$

then

$$\operatorname{tr} K = \int_{\mathbb{R}} \operatorname{tr}[K^*(t)K(t)]d\nu .$$

Proof. The operator $K = L^*L$, where $L:\tilde{\mathcal{H}} \to \mathcal{H}_\infty$ is given by $Lf = \int K(t)f(t)d\nu$ and $L^*g(t) = K^*(t)P(t)g$. Let $\{\varphi_i\}_{i=1}^{\infty}$ be a complete orthonormal system in $\mathcal{H}_\infty$. Then

$$\operatorname{tr} K = \sum_{i=1}^{\infty} \|L^*\varphi_i\|^2 = \sum_{i=1}^{\infty} \int \|K^*(t)P(t)\varphi_i\|^2 d\nu$$

$$= \int \sum_{i=1}^{\infty} \|K^*(t)P(t)\varphi_i\|^2 d\nu$$

$$= \int \operatorname{tr}[K^*(t)K(t)]d\nu .$$

The interchange of the orders of summation and integration follows from Fubini's Theorem. This ends the proof.

Proof of Theorem 4.2. Represent the operator $H = X + iY$ as in equation (3.22) of Theorem 3.1. The operator R^2 is given as the integral operator (3.23) on $\int_E \oplus \mathcal{H}_1(t)dt$. Directly from Lemma 4.1

$$\pi \operatorname{tr}[R^2] = \pi \operatorname{tr} Z = \frac{1}{2} \int_E \operatorname{tr}[B_1^2(t)]dt. \tag{4.4}$$

Obviously, $\frac{1}{2} \operatorname{tr}[B_1^2(t)] \leq \frac{1}{2}\|B_1^2(t)\| n_1(t) \leq \|Y\| n_1(t)$, almost everywhere on E. Inequality (4.3) follows. This completes the proof.

The identity (4.4) has the following corollary :

<u>Corollary 4.1</u>. Let $H = X + iY$ be a hyponormal operator which has a trace class self-commutator. In the singular integral representation of H, given by Theorem 3.1, the operators $B_1^2(t)$ $(t \in E)$ are trace class almost everywhere.

The next corollary of Theorem 4.2 appears in Putnam [11]. See also Herrero [1].

<u>Corollary 4.2</u>. Let $H = X + iY$ be hyponormal on $\mathcal{H}$ with $i[XY - YX] = R^2$. If the spectral multiplicity function of X restricte to the space $\mathcal{H}_1$ defined in (3.2) satisfies

$$n_1(t) < \infty \quad \text{a.e.,} \tag{4.5}$$

then R^2 is a compact operator.

Proof. Let $H = X + iY$ have the representation described in Theorem 3.1. If F is a measurable subset of E, then the projection o of $\int_E \oplus \mathcal{H}_1(t)dt$ onto $\int_F \oplus \mathcal{H}_1(t)dt$ will be simply denoted by F.

Let $E_j = \{t \in E : n_1(t) = j\}$, $E^j = \bigcup_{i=1}^{j} E_i$ and $F^j = E \setminus E^j$ $(j=1,2,\ldots)$. The hypothesis (4.5) implies $\text{meas}_1(F^j) \to 0$ as $j \to \infty$. By considering the operator $\tilde{H}$ compressed to $\int_{F^j} \oplus \mathcal{H}_1(t)dt$, we obtain from (4.2)

$$\pi \|F^j R^2 F^j\| \leq \text{meas}_1(F^j)\|Y\| .$$

Therefore, $F^j R^2 F^j$ converges to the zero operator as $j \to \infty$.

From Theorem 4.2, we know that $E^j R^2 E^j$ $(j = 1,2,\ldots)$ is compact. Further,

$$R^2 - E^j R^2 E^j = E^j R^2 F^j + F^j R^2 E^j + F^j R^2 F^j$$

and the three operators on the right all converge to zero. We conclude R^2 is compact. This ends the proof.

Notes

Section 1. For a fuller discussion of solutions of commutator equations the reader is referred to Kato [1 ,2]. The Friedrichs Γ-operators were defined by Friedrichs [1] in the study of self-adjoint perturbation problems. In the case where the self-adjoint commutator i[XY - YX] is a trace class operator (not necessarily semi-definite) the existence of $S_X^{\pm}$ and $\Gamma_X^{\pm}$, along with the representation of $S_X^{\pm}$ as singular integral operators, is developed in Carey and Pincus [5].

Section 2. The standard references for direct integral theory are the books of Dixmier [1] and Schwartz [1]. It is usually quite easy to reformulate any form of the spectral multiplicity theorem to obtain the version in Theorem 2.1.

Section 3. The first reseracher to obtain a singular integral representation for "nearly normal operators" was Xa Dao-xeng [1]. The vector valued generalization of Xa Dao-xeng's work was developed in Pincus [1] and Kato [2]. For the case of hyponormal operators, Muhly [1] mapped out the basic simplicity of this representation.

Section 4. The self-adjoint commutator inequality of Putnam [2], described in Theorem 4.1, was obtained by Putnam without the aid of the singular integral representation for hyponormal operators. The result in Theorem 4.1 plays an important role in establishing Putnam's result which shows the spectrum of a non-normal seminormal operator has positive two-dimensional measure. This result will be deduced in the next chapter. Theorem 4.1 is in Kato [2] and Putnam [1].

CHAPTER 3

TWO BASIC RESULTS ON SEMINORMAL OPERATORS

This chapter contains proofs of two of the deeper results on seminormal operators. The first of these is a result of Putnam [5] which establishes that the planar Lebesgue measure of the spectrum of a nonnormal seminormal operator is positive. The second is a result of Berger and Shaw [1] that shows every hyponormal operator with a cyclic vector has a trace class self-commutator. Actually, these results are related; however, here they are treated separately. The result of Berger and Shaw is developed below to serve as a point of departure for the study of seminormal operators with trace class self-commutator. This study begins in Chapter 4.

1. Cut Downs.

Let $S = X + iY$ be the Cartesian decomposition of the seminormal operator S on the Hilbert space $\mathcal{H}$. Assume $X = \int \lambda dE(\lambda)$ is the spectral resolution of X. Let β be a Borel set in the real line. The Hilbert space $\mathcal{H}_\beta$ is defined as the subspace $E(\beta)\mathcal{H}$ in $\mathcal{H}$. If $A \in L(\mathcal{H})$, then A_β will denote the operator $E(\beta)AE(\beta)$ acting on $\mathcal{H}_\beta$. The operator A_β will be referred to as the <u>cut down of A to the set β</u>. The cut down S_β of the seminormal operator S is seminormal. In fact, it is clear that

$$[S_\beta^*, S_\beta] = 2i[X_\beta Y_\beta - Y_\beta X_\beta] = [S^*, S]_\beta .$$

The following result (and proof) appears in Howe [1].

<u>Lemma 1.1</u>. Let $S = X + iY$ be a pure seminormal operator. Let $X = \int \lambda dE(\lambda)$ be the spectral resolution of X and β a Borel set in $\mathbb{R}$ such that $E(\beta) \neq 0$. The cut down $S_\beta = E(\beta)SE(\beta)$ to $\mathcal{H}_\beta = E(\beta)\mathcal{H}$ is pure.

Proof. Without loss of generality it can be assumed that the operator S is the hyponormal operator $H = X + iY$.

Suppose $\mathcal{H}_1 \subseteq \mathcal{H}_\beta$ is a subspace reducing H_β on which H_β is a normal operator. Set $\mathcal{H}_2 = \mathcal{H}_\beta \cap \mathcal{H}_1^\perp$ and $\mathcal{H}_3 = \mathcal{H}_\beta^\perp$. Relative to the orthogonal decomposition $\mathcal{H} = \mathcal{H}_1 \oplus \mathcal{H}_2 \oplus \mathcal{H}_3$ the operators X and Y have the matrix representations

$$X = \begin{bmatrix} X_1 & 0 & 0 \\ 0 & X_2 & 0 \\ 0 & 0 & X_3 \end{bmatrix} \qquad Y = \begin{bmatrix} Y_1 & 0 & R \\ 0 & Y_2 & S \\ R^* & S^* & Y_3 \end{bmatrix}.$$

Using the fact that $[X_1,Y_1] = 0$, a direct computation yields

$$i[X,Y] = i\begin{bmatrix} 0 & 0 & X_1R-RX_3 \\ 0 & [X_2,Y_2] & X_2S-SX_3 \\ X_3R^*-R^*X_1 & X_3S^*-S^*X_2 & [X_3,Y_3] \end{bmatrix}.$$

By considering vectors of the form $f = \lambda f_1 \oplus 0 \oplus f_3 \in \mathcal{H}_1 \oplus \mathcal{H}_2 \oplus \mathcal{H}_3$ $(\lambda \in \mathbb{R})$, we learn that the operator represented by

$$i\begin{bmatrix} 0 & 0 & X_1R-RX_3 \\ 0 & 0 & 0 \\ X_3R^*-R^*X_1 & 0 & 0 \end{bmatrix}$$

is non-negative. This can only happen when $X_1R = RX_3$. Since X_1 and X_3 are self-adjoint operators with disjoint support sets, then $R = 0$. In this case the subspace $\mathcal{H}_1$ reduces X and Y, consequently, $\mathcal{H}_1$ reduces H. The purity of H implies $\mathcal{H}_1 = \{0\}$. This completes the proof.

We are particularly interested in the cut down of seminormal operators to subintervals in $\mathbb{R}$. The main concern is the relation between the spectra of the operator and its cut down. The results will be stated in terms of the fine structure of the spectra.

For the remainder of this section $H = X + iY$ will denote a hyponormal operator. For simplicity of the discussion, the operator H will be assumed to be pure. Let $X = \int \lambda dE(\lambda)$ denote the spectral resolution of X. The purity of H implies X is an absolutely continuous operator (see, Theorem 3.2 of Chapter 2). If ϕ is any bounded Lebesgue measurable function, then $\phi(X)$ will denote the operator $\phi(X) = \int \phi(\lambda) dE(\lambda)$. With ϕ a fixed, bounded real valued measurable function, we define the operator

$$H' = X + i\phi(X)Y\phi(X) \quad . \tag{1.1}$$

he operator H' is hyponormal. A direct computation establishes

$$[(H')^*, H'] = \phi(X)[H^*, H]\phi(X) \quad . \tag{1.2}$$

Lemma 1.2. Let $\Delta = (a,b)$ be an interval in $\mathbb{R}$. Assume the function ϕ is a real valued, bounded measurable function which equals one on Δ. A complex number $z = x + iy$, with $x \in \Delta$, belongs to $\pi(H)$ if and only if z is in $\pi(H')$.

Proof. Assume first $z \in \pi(H)$. Let $\{f_n\}_{n=1}^{\infty}$ be a sequence of unit vectors such that $H_z f_n \to 0$. As in the proof of Proposition 5.1 in Chapter 1, $X_x f_n \to 0$ and $Y_y f_n \to 0$. Note for some $\delta > 0$,

$$\delta^2 \|E(\mathbb{R}\setminus\Delta) f_n\|^2 \leq \int_{\mathbb{R}\setminus\Delta} |\lambda - x|^2 d\|E(\lambda) f_n\|^2 \leq \|X_x f_n\|^2 \ .$$

Consequently, $\|E(\mathbb{R}\setminus\Delta) f_n\| \to 0$ and, therefore,

$$\|E(\Delta) f_n\| \to 1 \ , \quad X_x E(\Delta) f_n \to 0 \quad , \quad Y_y E(\Delta) f_n \to 0 \ . \tag{1.3}$$

Since, $Y_y E(\Delta) f_n = [Y\phi(X) - y] E(\Delta) f_n$, then

$$[\phi(X) Y \phi(X) - y] E(\Delta) f_n \to 0 \ . \tag{1.4}$$

Equations (1.3) and (1.4) imply $H_z' E(\Delta) f_n \to 0$ and, hence, $z \in \pi(H')$.

Next suppose $z \in \pi(H')$. For some sequence of unit vectors $\{f_n\}_{n=1}^{\infty}$, we have $H_z' f_n \to 0$. As above, $X_x f_n \to 0$ and $E(\mathbb{R}\setminus\Delta) f_n \to 0$.

Clearly $[(H')^*, H'] f_n \to 0$. From (1.2) we obtain $\phi(X)[H^*,H]\phi(X) f_n \to 0$. Therefore, $[H^*,H] E(\Delta) f_n \to 0$. From the identity

$$[H^*,H] E(\Delta) f_n = 2i[X_x Y_y - Y_y X_x] E(\Delta) f_n,$$

and the fact that $X_x E(\Delta) f_n \to 0$, we conclude $X_x Y_y E(\Delta) f_n \to 0$.

Again this implies $E(\mathbb{R}\setminus\Delta) Y_y E(\Delta) f_n \to 0$. Since, $E(\Delta) Y_y E(\Delta) f_n = \phi(X) E(\Delta) Y_y \phi(X) E(\Delta) f_n \to 0$, then $Y_y E(\Delta) f_n \to 0$. This last remark, coupled with the facts that $X_x E(\Delta) f_n \to 0$ and $\|E(\Delta) f_n\| \to 1$, implies $z \in \pi(H)$. This completes the proof.

<u>Corollary 1.1</u>. Let $\Delta = (a,b)$ and $z = x + iy$ satisfy $x \in \Delta$. Then $z \in \pi(H)$ if and only if $z \in \pi(H_\Delta)$.

Proof. Let ϕ be the characteristic function of Δ and H' the operator defined by (1.1) with this choice of ϕ. The operator H' is the orthogonal direct sum of the cut down H_Δ and the operator $X_{\mathbb{R}\setminus\Delta}$. The corollary clearly follows from Lemma 1.2. This ends the proof.

If T is an operator on the Hilbert space $\mathcal{H}$, then the notation $\Sigma(T)$ will be employed for the <u>residual spectrum</u> of the operator T. Thus $z \in \Sigma(T)$ if and only if there is an $\varepsilon > 0$ such that

$$\|(T-z)f\| \geq \varepsilon\|f\|$$

for every $f \in \mathcal{H}$; however, T-z is not onto. The set $\Sigma(T)$ is open and the spectrum may be written as the disjoint union

$$\sigma(T) = \pi(T) \cup \Sigma(T) . \tag{1.5}$$

Lemma 1.3. Let $\Delta = (a,b)$ and $z = x + iy$ satisfy $x \in \Delta$. The number z belongs to $\Sigma(H)$ if and only if z belongs to $\Sigma(H_\Delta)$.

Proof. The result will be established by showing that $z \in \rho(H)$ if and only if $z \in \rho(H_\Delta)$. This in conjunction with Lemma 1.2 and identity (1.5) implies the desired result.

For $0 \leq t \leq 1$, define the functions ϕ_t on the real line by

$$\phi_t(\lambda) = \begin{cases} 1 - t & \lambda \notin \Delta \\ 1 & \lambda \in \Delta \end{cases}$$

and the hyponormal operators

$$H_t = X + i\phi_t(X)Y\phi_t(X).$$

The mapping $t \to H_t$ from $[0,1]$ into $L(\mathcal{H})$ is continuous in the operator norm. Note that $H_0 = H$ and $H_1 = X_{\mathbb{R}\setminus\Delta} \oplus H_\Delta$. It suffices to show $z \in \rho(H_0)$ if and only if $z \in \rho(H_1)$.

First assume $z \in \rho(H_0)$. There is a disc $\mathbb{D} \subset \rho(H_0)$ of radius $\delta > 0$ centered at z with $\mathbb{D} \subseteq (a,b) \times \mathbb{R}$. By virtue of Lemma 1.1, we know

$$\pi(H_t) \cap \mathbb{D} = \emptyset, \quad t \in [0,1] . \tag{1.6}$$

Suppose $z \in \sigma(H_1)$. Let s be defined by

$$s = \inf\{t \in [0,1]: z \in \sigma(H_t)\}. \tag{1.7}$$

By the assumption that $z \in \rho(H_0)$ and the norm continuity of the mapping $t \to H_t$, we have $0 < s \leq 1$.

Choose an increasing sequence $\{s_n\}_{n=1}^\infty \subset [0,1]$ with $s_n \to s$. From (1.6) (using the fact that $\Sigma(T)$ is open) we conclude $\mathbb{D} \subset \rho(H_{s_n})$ $(n = 1,2,\ldots)$. By Corollary 1.1 of Chapter 1,

$$\|(H_{s_n} - z)^* f\| \geq \delta\|f\|, \quad f \in \mathcal{H}, \ (n=1,2,\ldots).$$

This implies

$$\|(H_s - z)^* f\| \geq \delta\|f\|, \quad f \in \mathcal{H} .$$

Consequently, $z \notin \sigma(H_s)$. However, if $z \notin \sigma(H_s)$, then $z \in \rho(H_t)$ when t is close to s. This contradicts (1.7). We conclude $z \in \rho(H_1)$.

Next suppose $z \in \rho(H_1)$. The argument to establish $z \in \rho(H_0)$ is similar to the argument above. We sketch the modifications. There is a disc $\mathbb{D} \subset \rho(H_1)$ of radius $\delta > 0$ centered at z with $\mathbb{D} \subset (a,b) \times \mathbb{R}$. Again by Lemma 1.1, we know

$$\pi(H_t) \cap \mathbb{D} = \emptyset, \quad t \in [0,1].$$

Suppose $z \in \sigma(H_0)$. Let s be defined by

$$s = \sup\{t \in [0,1]: z \in \sigma(H_t)\}.$$

This time, we have $0 \leq s < 1$. Choose a decreasing sequence $\{s_n\}_{n=1}^{\infty} \subset [0,1]$ such that $s_n \to s$. Arguing as above, we deduce, $z \notin \sigma(H_1)$. This completes the proof.

<u>Corollary 1.2</u>. Let $\Delta = (a,b)$ be an open interval in $\mathbb{R}$. Then

$$\sigma(H_\Delta) \cap (\Delta \times \mathbb{R}) = \sigma(H) \cap (\Delta \times \mathbb{R}) . \tag{1.8}$$

This corollary follows immediately from Corollary 1.1 and Lemma 1.3.

2. Putnam's Inequality.

In this section we derive one of the deepest results about seminormal operators. This result of Putnam asserts that the planar Lebesgue measure of the spectrum of a non-normal seminormal operator is positive. The precise result is the following theorem in Putnam [5]

<u>Theorem 2.1</u>. Let S be a seminormal operator on a Hilbert space $\mathcal{H}$ with self-commutator $D = S^*S - SS^*$. Then

$$\pi\|D\| \leq \text{meas}_2(\sigma(S)), \qquad (2.1)$$

where meas_2 denotes planar Lebesgue measure.

Proof. Without loss of generality it can be assumed that $S = H = X + iY$ is a pure hyponormal operator. The operator X will be assumed to have the spectral resolution $X = \int \lambda dE(\lambda)$.

Let $\Delta = [a,b)$ be an interval in $\mathbb{R}$ and $c = \frac{a+b}{2}$ the center of Δ. Applying Theorem 4.1 of the preceding chapter to the pair $-Y_\Delta, (X-c)_\Delta$, we obtain for $f \in \mathcal{H}$,

$$\pi(E(\Delta)DE(\Delta)f,f) = \pi\|D^{1/2}E(\Delta)f\|^2 \leq |\Delta|\,\text{meas}_1(\sigma(Y_\Delta))\|E(\Delta)f\|^2. \qquad (2.2)$$

Note we have used $2\|(X-c)_\Delta\| \leq |\Delta|$, with $|\Delta|$ denoting the length of Δ.

We introduce the following notation

$$Q(\Delta) = \text{meas}_1(\text{proj}_y(\sigma(H_\Delta))) .$$

After Proposition 5.1 of Chapter 1, we have

$$\text{meas}_1[\sigma(Y_\Delta)] = \text{meas}_1(\text{proj}_y(\sigma(H_\Delta))) .$$

Substituting our new notation into (2.2), we obtain

$$\pi^{1/2}\|D^{1/2}E(\Delta)f\| \leq |\Delta|^{1/2}[Q(\Delta)]^{1/2}\|E(\Delta)f\| . \qquad (2.3)$$

Let $[\alpha,\beta]$ be an interval in $\mathbb{R}$ which contains the spectrum of the operator X and let

$$P: \alpha = c_0 < c_1 < \ldots < c_n = \beta$$

be a partition of $[\alpha,\beta]$. Set $\Delta_i = [c_{i-1},c_i)$, $i = 1,2,\ldots,n$. For $f \in \mathcal{H}$,

$$\pi^{1/2}\|D^{1/2}f\| \leq \pi^{1/2}\sum_{i=1}^{n}\|D^{1/2}E(\Delta_i)f\|$$

$$\leq \sum_{i=1}^{n}|\Delta_i|^{1/2}[Q(\Delta_i)]^{1/2}\|E(\Delta_i)f\| \qquad (2.4)$$

$$\leq \{\sum_{i=1}^{n} Q(\Delta_i)\ |\Delta_i|\}^{1/2}\|f\| ,$$

where the second inequality follows from (2.3).

Let F be the function on $[\alpha,\beta]$ whose value on Δ_i is $Q(\Delta_i)$, $i = 1,\dots,n$. Inequality (2.4) may be rewritten in the form

$$\pi(Df,f) \leq (\int_\alpha^\beta F(x)dx)\|f\|^2. \qquad (2.5)$$

Select a sequence of partitions $\{P_n\}_{n=1}^{\infty}$ of $[\alpha,\beta]$, with P_{n+1} a refinement of P_n, and such that the norm of P_n tends to zero as $n \to \infty$. Let F_n be the function corresponding to P_n in the same manner that the function F corresponded to the partition P above.

Suppose $\Delta = [\gamma,\delta)$ is a subinterval formed from consecutive points in P_{n+1} which is contained in the interval $\Delta' = [\gamma',\delta')$ formed from consecutive points in the partition P_n. Directly from Corollary 1.2 we conclude

$$\sigma(H_\Delta) \cap (\Delta \times \mathbb{R}) = \sigma(H_{\Delta'}) \cap (\Delta \times \mathbb{R}) .$$

As a consequence, the only difference between the spectrum of H_Δ and $\sigma(H_{\Delta'}) \cap (\overline{\Delta} \times \mathbb{R})$ is possibly exposed subsets of line segments containe in $\mathrm{Re}z = \gamma$ or $\mathrm{Re}z = \delta$. (Here, we are using the notation $\overline{\Delta}$ for the closure of the interval Δ.) It follows from Lemma 1.1 that the operator H_Δ is pure; therefore, by Proposition 5.3 in Chapter 1, no such line segments can exist. This leads to the inclusion:

$$\sigma(H_\Delta) \subset \sigma(H_{\Delta'}) \cap (\overline{\Delta} \times \mathbb{R}) . \qquad (2.6)$$

In particular, $Q(\Delta) \leq Q(\Delta')$ and, hence, the sequence of functions $\{F_n\}$

satisfy $F_{n+1} \leq F_n$, $n = 1,2,\dots$.

With each partition P_n $(n = 1,2,\dots)$ we will also associate the function:

$$S_n(x) = \text{meas}_1\{\text{proj}_y[\sigma(H) \cap (\bar{\Delta} \times \mathbb{R})]\} ,$$

whenever x belongs to the interval Δ formed from consecutive points in P_n.

The argument used to establish (2.6) also shows $\sigma(H_\Delta) \subset \sigma(H) \cap (\bar{\Delta} \times \mathbb{R})$. This implies

$$F_n(x) \leq S_n(x),\ n = 1,2,\dots \ .$$

Inequality (2.5) holds with F replaced by F_n. Applying the monotone convergence theorem, we obtain

$$\pi(Df,f) \leq \lim_n \int_\alpha^\beta F_n(x)dx = \int_\alpha^\beta \lim_n F_n(x)dx \leq \int_\alpha^\beta \underline{\lim}\, S_n(x)dx. \quad (2.7)$$

It is easy to verify that $\lim S_n(x)$ exists, when $x \notin \cup P_n$, and equals

$$\text{meas}_1 \{\sigma(H) \cap (\{x\} \times \mathbb{R})\}.$$

Substituting in (2.7) we obtain (2.1). This completes the proof.

A compact subset K of the plane is said to have <u>positive (planar) density</u> in case, whenever a disc $\mathbb{D}$ intersects K, then $\text{meas}_2(\mathbb{D} \cap K)$ is positive.

<u>Corollary 2.1</u>. Let S be a pure seminormal operator. Then $\sigma(S)$ is a set of positive density.

Proof. Assume $\Delta = [a,b]$ and $\delta = [c,d]$ are intervals in $\mathbb{R}$ such that the interior of the rectangle $\Delta \times \delta$ intersects $\sigma(S)$.

Let $S = X + iY$ be the Cartesian form of S and $X = \int\lambda dE(\lambda)$ be the spectral resolution of X. Let $S_\Delta = X_\Delta + iY_\Delta$ be the cut down of S to $\mathcal{H}_\Delta = E(\Delta)\mathcal{H}$. Suppose $Y_\Delta = \int\lambda dF^\Delta(\lambda)$ is the spectral resolution of Y_Δ

and denote by $S_{\Delta,\delta}$ the operator $(S_\Delta)_\delta = F^\Delta(\delta)S_\Delta F^\Delta(\delta)$ on $\mathcal{H}_{\Delta,\delta} = F^\Delta(\delta)\mathcal{H}_\Delta$. The operator $S_{\Delta,\delta}$ is pure (Lemma 1.1) and from Proposition 5.1 of Chapter 1, $\sigma(S_{\Delta,\delta}) \subset \Delta \times \delta$.

Theorem 2.1 implies $\text{meas}_2\sigma(S_{\Delta,\delta})$ is positive. Two applications of Corollary 1.2 imply that the portion of $\sigma(S_{\Delta,\delta})$ in the interior of $\Delta \times \delta$ lies in $\sigma(S)$. This means $\sigma(S) \cap (\Delta \times \delta)$ has positive measure and completes the proof.

3. A Result of Berger and Shaw.

One of the most striking results in the theory of hyponormal operators is the result of Berger and Shaw [1] which establishes that a hyponormal operator with a cyclic vector has a trace class self-commutator. This result is derived below and is intended as motivation for the study of seminormal operators with trace class self-commutators. This study occupies the main portion of the remainder of these notes.

The results presented below are a very thin subset of the work of Berger and Shaw [1]. It is hoped they will serve to whet the reader's appetite for the remainder of the work of these authors. Also in Chapter 5, we will derive a more recent result of Berger [1] that connects the Pincus principal function of a hyponormal operator (this function is defined in Chapter 5) to the existence of cyclic vectors. The argument to deduce this new result of Berger is basically the same as the method described in this section. The reader will be at an advantage if we first outline the argument in the simpler case.

The following lemma appears in Berger and Shaw [1]. The proof of the lemma would be somewhat simpler if we considered only intertwining operators that belong to the trace class. Here we will expend a littl extra effort to handle the case of Hilbert-Schmidt intertwinings. We recall that an operator W on the Hilbert space $\mathcal{H}$ is said to be <u>Hilbert Schmidt</u> in case,

$$\sum_{i,j} |(W\varphi_i, \varphi_j)|^2 = \sum_i \|W\varphi_i\|^2$$

is finite for some (hence, for all) complete orthonormal system $\{\varphi_i\}$ in $\mathcal{H}$. The properties of Hilbert-Schmidt operators are described in Schatten [1] and Gohberg-Krein [1].

Lemma 3.1. Let T and H be hyponormal operators on a Hilbert space $\mathcal{H}$. Assume W is a one-to-one Hilbert-Schmidt operator with dense range satisfying

$$WT = HW \quad . \tag{3.1}$$

Then

$$\mathrm{tr}[H^*,H] \le \mathrm{tr}[T^*,T]. \tag{3.2}$$

Proof. Without loss of generality it can be assumed that $\mathrm{tr}[T^*,T]$ is finite. The proof of (3.2) is basically obtained by considering the hyponormal operator $T \oplus H$ restricted to the family of invariant subspaces

$$\mathcal{S}(t) = \{(x,tWx) : x \in \mathcal{H}\} \ , \ t \in \mathbb{R} \quad .$$

The subspaces $\mathcal{S}(t)$ are the graphs of the operators tW and by virtue of (3.1) are invariant under $T \oplus H$.

The orthogonal projection J_t of $\mathcal{H} \oplus \mathcal{H}$ onto $\mathcal{S}(t)$ has the 2×2 matrix form

$$J_t = \begin{bmatrix} \dfrac{1}{1+t^2W^*W} & \dfrac{1}{1+t^2W^*W}\, tW^* \\ \\ tW\dfrac{1}{1+t^2W^*W} & tW\dfrac{1}{1+t^2W^*W}\, tW^* \end{bmatrix} . \tag{3.3}$$

Things will be a little simpler, if instead of considering $T \oplus H$ restricted to $\mathcal{S}(t)$, we consider the family of hyponormal operators $A_t = [T \oplus H]J_t$ $(t \in \mathbb{R})$ acting on $\mathcal{H} \oplus \mathcal{H}$.

It will first be shown that for any $t \in \mathbb{R}$

$$\mathrm{tr}[A_t^*,A_t] = \mathrm{tr}[T^*,T] \quad . \tag{3.4}$$

It is easy to see that $J_t = L_t + Q_t$, where

$$L_t = \begin{bmatrix} I & \frac{1}{1+t^2W^*W}\, tW^* \\ tW\,\frac{1}{1+t^2W^*W} & 0 \end{bmatrix}$$

and Q_t is a trace class operator. Therefore,

$$[A_t^*, A_t] = [L_t(T \oplus H)^*, (T \oplus H)L_t] + R_t ,$$

where R_t is a trace class operator satisfying

$$\operatorname{tr} R_t = 0 .$$

Let Z_t be the Hilbert-Schmidt operator $(1+t^2W^*W)^{-1}tW^*$. A direct computation shows that the self-commutator $[L_t(T \oplus H)^*, (T \oplus H)L_t]$ has the form

$$\begin{bmatrix} [T^*,T] + Z_tH^*HZ_t^* - TZ_tZ_t^*T^* & T^*TZ_t - TZ_tH^* \\ Z_t^*T^*T - HZ_t^*T^* & Z_t^*T^*TZ_t - HZ_t^*Z_tH^* \end{bmatrix} .$$

The 1,2-entry in the above matrix may be rewritten in the form

$$\begin{aligned} T^*TZ_t - TZ_tH^* &= T^*TZ_t - T(1+t^2W^*W)^{-1}T^*(tW^*) \\ &= [T^*T - TT^*]Z_t + S_t , \end{aligned}$$

where S_t is a trace class operator. This last identity shows that the 1,2-entry (hence, the 2,1-entry) is trace class. Now it is clear that $[A_t^*, A_t]$ is trace class and

$$\begin{aligned}
\mathrm{tr}[A_t^*, A_t] &= \mathrm{tr}[L_t(T \oplus H)^*, (T \oplus H)L_t] \\
&= \mathrm{tr}[T^*T] + \mathrm{tr}(Z_tH^*HZ_t^* - \mathrm{tr}(TZ_tZ_t^*T^*) \\
&\quad + \mathrm{tr}(Z_t^*T^*TZ_t) - \mathrm{tr}(HZ_t^*Z_tH^*) \\
&= \mathrm{tr}[T^*T] .
\end{aligned}$$

This establishes (3.4).

We remark that to this point, we have not used the hypothesis that T and H are hyponormal operators or the assumption that W is one-to-one with dense range. This latter assumption implies

$$s - \lim J_t = \begin{bmatrix} 0 & 0 \\ 0 & I \end{bmatrix} .$$

If $\{z_n\}_{n=1}^{\infty}$ is any orthonormal system in $\mathcal{K} \oplus \mathcal{K}$, then

$$\begin{aligned}
\mathrm{tr}[T^*, T] &= \mathrm{tr}[A_t^*, A_t] \\
&\geq \sum_{n=1}^{\infty} [\|(T \oplus H)J_tz_n\|^2 - \|J_t[T \oplus H]^*J_tz_n\|^2] . \quad (3.5)
\end{aligned}$$

Let $z_n = 0 \oplus \phi_n$, where $\{\phi_n\}_{n=1}^{\infty}$ is a complete orthonormal system in $\mathcal{K}$. The term

$$\|(T \oplus H)J_tz_n\|^2 - \|J_t[T \oplus H]^*J_tz_n\|^2$$

appearing in (3.5) is non-negative and converges to

$$\|H\phi_n\|^2 - \|H^*\phi_n\|^2$$

as $t \to \infty$. Applying Fatou's lemma in (3.5), we obtain $\mathrm{tr}[H^*, H] \leq \mathrm{tr}[T^*, T]$. This completes the proof.

The following theorem is a simple version of the main result of Berger and Shaw [1].

Theorem 3.1. Let H be a hyponormal operator on a Hilbert space $\mathcal{H}$. Suppose the operator H possesses a cyclic vector. Then the self-commutator $[H^*,H]$ is trace class.

Proof. Without loss of generality it can be assumed $\|H\| < 1$. Let f be a cyclic vector for H. Define $W:\ell_2^+ \to \mathcal{H}$ by

$$W\{a_i\}_{i=0}^{\infty} = \sum_{i=0}^{\infty} a_i H^i f. \tag{3.6}$$

The operator W is Hilbert-Schmidt. In fact, let $\{e_n\}_{n=0}^{\infty}$ be the usual basis in ℓ_2^+. Then

$$\sum_{n=0}^{\infty} \|We_n\|^2 = \sum_{n=0}^{\infty} \|H^n f\|^2 \leq \left(\sum_{n=0}^{\infty} \|H\|^{2n}\right) \|f\|^2 < \infty .$$

Let U_+ be the unilateral shift on ℓ_2^+. From (3.6), we know

$$WU_+ = HW .$$

The hypothesis that f is a cyclic vector for H guarantees that W has dense range. If $W(\{a_i\}) = 0$, then the operator $\sum_{i=0}^{\infty} a_i H^i$ would be the zero operator. The only way this could be possible, is if the spectrum of H is a finite set. We can ignore this case. Therefore, W may be assumed to be one-to-one with dense range.

Immediately from Lemma 3.1, we conclude

$$\operatorname{tr}[H^*,H] \leq \operatorname{tr}[U_+^*,U_+] = 1 .$$

This completes the proof.

Remark. The operator W constructed in (3.6) for the proof of Theorem 3.1 is actually trace class. This can be seen as follows. Let $\|H\| < r < 1$. The operator W factors in the form $W = W_1 W_2$, where $W_2:\ell_2^+ \to \ell_2^+$ is defined by

$$W_2\{a_i\}_{i=0}^{\infty} = \{a_i r^i\}_{i=0}^{\infty}$$

and $W_1 : \ell_2^+ \to \mathcal{K}$ is given by

$$W_1\{a_i\}_{i=0}^{\infty} = \sum_{i=0}^{\infty} a_i \left(\frac{H}{r}\right)^i f \ .$$

The operators W_1 and W_2 are Hilbert-Schmidt and, therefore, W is a trace class operator. Later, in Chapter 5, we will take advantage of the fact that the intertwining operator defined in (3.6) is trace class. As we stated earlier, when W is trace class, Lemma 3.1 is easier to prove. In fact, in this case

$$J_t - \begin{bmatrix} I & 0 \\ 0 & 0 \end{bmatrix}$$

is trace class and (3.4) is immediate.

Notes

Section 1. Putnam was the first to study cut downs of seminormal operators. Indeed, the cut down concept plays a basic role in many of Putnam's investigations. Lemma 1.1 is established in Howe [1] and also in Putnam [5]. Lemma 1.2 and Corollary 1.1 are basically in Putnam [4]. The proof of Lemma 1.3 is an adaptation of the arguments in Howe [1] and Clancey [1]. The result in Lemma 1.3, which is the key to Putnam's spectral inequality (Theorem 2.1), was obtained by Putnam in [5] and [7]. Howe [1] has given a nice treatment of cut downs.

Section 2. The inequality (2.1) was first obtained by Putnam [5]. In the case where the self-commutator $D = S^*S - SS^*$ is compact, see Clancey [1]. Berger and Shaw [1] have given a separate derivation of Putnam's inequality. The result on positive density (Corollary 2.1) is in Putnam [5]. Actually, the result in Corollary 2.1 characterizes the spectrum of pure seminormal operators. A necessary and sufficient condition that a compact set K be the spectrum of a pure seminormal operator is that K have positive density. This result is in Putnam [9] and also Carey and Pincus [4].

Section 3. All of the results in this section are simple cases of the results in Berger and Shaw [1].

CHAPTER 4

PERTURBATION DETERMINANTS AND THE PHASE SHIFT

The work of Berger and Shaw, which was briefly discussed in the last chapter, throws the spotlight on the study of seminormal operators with trace class self-commutators. Pincus [1] has associated a scalar principal function with every operator having a trace class self-commutator. This function and some of its basic properties will be developed in the next chapter. It is important to realize that the principal function is a natural two-variable generalization of a function, called the phase shift, which has been studied in connection with trace class perturbations of self-adjoint operators. For this reason, it is illustrative to highlight some of the theory connected ith the phase shift. This is our concern in this short expository chapter.

. Infinite Determinants.

In the present section we briefly recall the definition and basic roperties of infinite determinants. No proofs are given. The reader s referred to Gohberg and Krein [1] for further details.

Let T be a compact operator on the (separable) infinite dimensionl Hilbert space $\mathcal{H}$. Let $T = WK$ be the polar factorization of T. Then $= (T^*T)^{1/2}$ and W is a partial isometry with initial space $[\ker T]^\perp$ nd final space $\overline{R(T)}$. The s-numbers $\{s_j(T)\}_{j=1}^{\infty}$ of the operator T are he eigenvalues of K, counted with respect to multiplicity and arranged in decreasing order.

A compact operator T on $\mathcal{H}$ is said to be in the trace class in ase the trace norm:

$$\|T\|_1 = \sum_{j=1}^{\infty} s_j(T)$$

is finite. Let T be a trace class operator. The notation:

$$\Lambda(T) = \{\lambda_j(T) : j = 1,2,\ldots,\nu(T)\}$$

will denote an enumeration of the non-zero eigenvalues of T counted with respect to algebraic multiplicities. The infinite determinant of the operator I + T is defined by the formula

$$\det(I+T) = \begin{cases} \prod_{j=1}^{\nu(T)} (1+\lambda_j(T)), & \Lambda(T) \neq \emptyset \\ 1 & \Lambda(T) = \emptyset \end{cases} \quad (1.1)$$

In case $\nu(T)$ is infinite, the convergence of the product in (1.1) follows from the inequality

$$\sum_{j=1}^{\nu(T)} |\lambda_j(T)| \leq \|T\|_1 \quad .$$

This last inequality appears, for example, in Gohberg-Krein [1, Chapter II].

Some of the properties of the infinite determinant which will be used in the sequel are the following:

1°. The determinant is continuous with respect to the trace norm. This means that if $\{T_n\}_{n=1}^{\infty}$ is a sequence of trace class operators converging to the operator T in the norm $\| \ \|_1$, then

$$\lim_{n\to\infty} \det(I+T_n) = \det(I+T) \quad .$$

2°. Let T be trace class and let $\{\phi_i\}_{i=1}^{\infty}$ be a complete orthonormal set in $\mathcal{H}$. For n = 1,2,..., define the n×n matrix

$$[\delta_{ij} + (T\phi_i,\phi_j)]_{n\times n} \quad .$$

Then

$$\det(I+T) = \lim_{n\to\infty} \det[\delta_{ij} + (T\phi_i,\phi_j)]_{n\times n} . \qquad (1.2)$$

This result is a consequence of 1° and the following fact: If T is trace class and $\{P_n\}_{n=1}^{\infty}$ is a sequence of orthogonal projectons converging strongly to the identity, then $\|P_nTP_n - T\|_1 \to 0$.

Equation (1.2) permits us to lift many of the familiar algebraic properties of determinants for matrices to the case of infinite determinants. For example

3°. Let T be trace class and $A \in L(\mathcal{H})$. Then

$$\det(I+TA) = \det(I+AT) .$$

4°. Let T_1 and T_2 be trace class. Then

$$\det((I+T_1)(I+T_2)) = \det(I+T_1)\det(I+T_2) .$$

5°. Let T be a trace class operator with $\|T\|_1 < 1$. Then

$$\det(I+T) = \exp \operatorname{tr}(\log(I+T)) .$$

ere log(I+T) is the logarithm of I+T given by the series (convergent n the norm $\| \ \|_1$)

$$\log(I+T) = -\sum_{n=1}^{\infty} \frac{(-1)^n T^n}{n} .$$

Finally, we record the following:

6°. Let $\lambda \to T(\lambda)$ be an analytic trace class valued function efined on an open subset Ω of the complex plane. This means that (λ) may be expanded in a $\| \ \|_1$-convergent power series in a neighborood of each point in Ω. The function

$$\det(I + T(\lambda))$$

s analytic on Ω.

2. Perturbation Determinants.

Let A be a self-adjoint operator and D a self-adjoint, trace class operator on $\mathcal{H}$. The perturbed operator A+D will be denoted by B. The perturbation determinant corresponding to the perturbation $A \to B = A+D$ is the function

$$\Delta_{B/A}(z) = \det((B-z)(A-z)^{-1})$$
$$= \det(I + D(A-z)^{-1}) , \quad z \notin \sigma(A) . \qquad (2.1)$$

After property 6° of the preceding section, it is clear that $\Delta_{B/A}$ defines an analytic function on the complement of $\sigma(A)$.

The perturbation determinant has the following important multiplicative property. Let A, B, C be self-adjoint operators such that $D_1 = B-A$ and $D_2 = C-B$ are trace class operators. Then

$$\Delta_{C/A}(z) = \Delta_{C/B}(z)\Delta_{B/A}(z) , \quad z \notin \sigma(A) \cup \sigma(B) . \qquad (2.2)$$

The identity (2.2) follows from property 4° of Section 1 and the relation

$$(C-z)(A-z)^{-1} = (C-z)(B-z)^{-1}(B-z)(A-z)^{-1}$$
$$= (I+D_2(B-z)^{-1})(I+D_1(A-z)^{-1}) .$$

One of the basic results on perturbation determinants is an exponential representation of Krein [1]. This representation and some of its basic properties will be derived below. We consider first the case of one-dimensional perturbations.

Let A be a self-adjoint operator on $\mathcal{H}$ and let D be the rank one operator $Df = \lambda(f,k)k$, where $\lambda \in \mathbb{R}$ and k is a unit vector in $\mathcal{H}$. Set $B = A+D$. The perturbation determinant has the form

$$\Delta_{B/A}(z) = 1 + \lambda((A-z)^{-1}k,k), \ \text{Im}\ z \neq 0 . \qquad (2.3)$$

The cases $\lambda > 0$ and $\lambda < 0$ will be considered separately:

If $\lambda > 0$, then (2.3) makes it obvious that $\Delta_{B/A}$ is an analytic function in the upper half-plane with non-negative imaginary part. In fact, if $A = \int t dE(t)$ denotes the spectral resolution of A, then

$$\Delta_{B/A}(z) = 1 + \int \frac{1}{t-z} d\mu(t), \qquad (2.4)$$

where $d\mu(t) = \lambda d\|E(t)k\|^2$ is a non-negative measure on $\mathbb{R}$ having compact support.

A theorem of Verblunsky (See, e.g. Aronszajn and Donoghue [1]) provides the representation

$$\Delta_{B/A}(z) = \exp \int \frac{\eta(t)}{t-z} dt , \qquad (2.5)$$

where η is a measurable function on $\mathbb{R}$ satisfying $0 \le \eta \le 1$. The function η providing the representation (2.5) is called the <u>phase shift</u> of the perturbation $A \to A+D$.

We will not go into a detailed proof of Verblunsky's Theorem; however, the idea is the following: Let F be the function appearing in the right side in (2.4). Thus

$$F(z) = 1 + \int \frac{d\mu(t)}{t-z} .$$

Then $G = \log F$ is an analytic function in the upper half-plane with a non-negative imaginary part. (The logarithm function is the principal branch: $-\pi < \mathrm{Im}(\log z) \le \pi$.) Consequently, by a theorem of Nevanlinna (See, e.g. Duren [1]) there is a non-negative measure ν on $\mathbb{R}$ providing the representation

$$G(z) = \alpha z + \beta + \int \frac{1+tz}{1+t^2} \frac{d\nu(t)}{t-z} ,$$

where α, β are complex numbers and $\int (t^2+1)^{-1} d\nu < \infty$. Taking into account that $0 \le \mathrm{Im}\, G \le \pi$, it is routine to establish the existence and boundedness of $\frac{d\nu}{dt}$. The behaviour of G at infinity allows G to be rewritten in the form

$$G(z) = \int \frac{\eta(t)}{t-z}\, dt \quad .$$

This provides the representation (2.5).

A direct computation establishes

$$1 + \lambda((A-z)^{-1}k,k) = \{1 - \lambda((A+D-z)^{-1}k,k)\}^{-1} \quad .$$

Consequently, when λ is negative, we have the representation

$$\Delta_{B/A}(z) = \exp \int \frac{\eta(t)}{t-z}\, dt, \tag{2.6}$$

where $-1 \leq \eta \leq 0$ and $-\eta$ is the phase shift of the (non-negative) perturbation $A+D \to A$.

The relationship between the measure μ and the phase shift η has been studied in considerable detail by Aronszajn and Donoghue [1]. We record some of the salient properties of this relationship.

We note first that the perturbation determinant and the phase shift depend only on the operator A restricted to the <u>interaction space</u> $H(A,D)$. This space is

$$H(A,D) = \bigvee \{A^j k:\ k \in R(D),\ j = 0,1,2,\dots\} \quad .$$

In view of the simple relation between the cases $\lambda > 0$ and $\lambda < 0$, we confine our remarks to the case $\lambda > 0$. The following two properties of the phase shift are easily verified:

1°. The function η is determined almost everywhere by

$$\eta(t) = \lim_{y \downarrow 0} \frac{1}{\pi} \arg(\Delta_{B/A}(t+iy)), \tag{2.7}$$

where $\arg(z)$ $(z \neq 0)$ denotes the principal value of the argument chosen such that $-\pi < \arg(z) \leq \pi$.

2°. The function η is supported on the interval

$$[m(A,D),\ M(A,D) + \|D\|],$$

where $$m(A,D) = \inf\{(Ax,x): \|x\| = 1,\ x \in H(A,D)\}\ ,$$
$$M(A,D) = \sup\{(Ax,x): \|x\| = 1,\ x \in H(A,D)\}\ .$$

Before recording further properties of the phase shift it will be instructive to consider an example.

Let A be a self-adjoint $n \times n$ matrix considered as an operator on $\mathbb{C}^n$. Let $Df = (f,y)y$ be a one dimensional operator on $\mathbb{C}^n$ with range spanned by the vector $y \in \mathbb{C}^n$. After diagonalization of A the phase shift of the perturbation $A \to A+D$ has a fairly explicit description. In fact, assume A is diagonalized as the matrix

$$\hat{A} = \begin{pmatrix} \alpha_{11} & & & \\ & \cdot & & 0 \\ & & \cdot & \\ 0 & & & \cdot \\ & & & \alpha_{nn} \end{pmatrix}$$

here $\alpha_{11} \leq \alpha_{22} \leq \cdots \leq \alpha_{nn}$. In this diagonalization of A the operaor $D = (\ ,y)y$ is unitarily equivalent to $\tilde{D} = (\ ,x)x$, for some $= (x_1,\ldots,x_n) \in \mathbb{C}^n$.

For z not equal to any of the eigenvalues $(\alpha_{11},\ldots,\alpha_{nn})$, the erturbation determinant has the form

$$\Delta_{B/A}(z) = 1 + (\alpha_{11}-z)^{-1}|x_1|^2 + \ldots + (\alpha_{nn}-z)^{-1}|x_n|^2\ . \quad (2.8)$$

rom (2.7) it is clear that the phase shift for the perturbation $\to A+D$ equals the characteristic function of the set in $\mathbb{R}$ where the unction (2.8) is negative. The graph of (2.8) on $\mathbb{R}$ is the following:

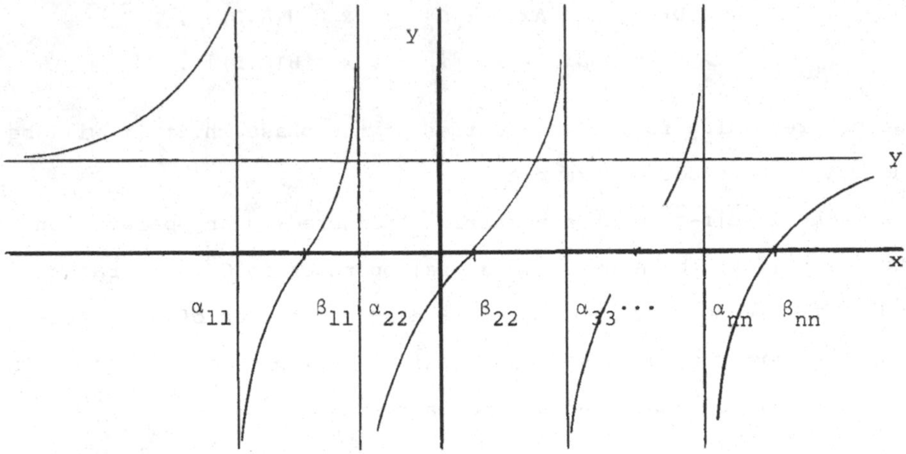

Therefore, η is the characteristic function of the set

$$(\alpha_{11},\beta_{11}) \cup (\alpha_{22},\beta_{22}) \cup \cdots \cup (\alpha_{nn},\beta_{nn}) .$$

The exact values of $\beta_{11},\ldots,\beta_{nn}$ are hard to come by.

We continue with our list of properties of the phase shift for one-dimensional perturbations. The next three properties of the phase shift are established in Aronszajn and Donoghue [1]. These properties will not be used below. The decision to list these specific properties is made on the basis of the important role which they play in related investigations. (See, e.g. Carey and Pincus [1,7].)

3°. The operator A restricted to H(A,D) is purely singular if and only if η equals a.e. the characteristic function of some subset of $\mathbb{R}$.

4°. A real number α is an eigenvalue of the operator A on H(A,D) (equivalently, the measure μ has a point mass at $t = \alpha$) if and only if

$$\int_{\alpha}^{1+\alpha} \frac{1-\eta(t)}{t-\alpha}\,dt + \int_{-1+\alpha}^{\alpha} \frac{\eta(t)}{\alpha-t}\,dt < \infty .$$

The reader can see easily how 3° and 4° are valid for the matrix example discussed above.

Finally, we state the following sufficient condition for the absolute continuity of the measure μ.

5°. Suppose $a < b$ and

$$\omega(\eta:a,b) = \sup\{|\eta(t) - \eta(t')| : t,t' \in (a,b)\}$$

satisfies

$$\omega(\eta:a,b) < 1.$$

Then the measure $d\eta(t) = d\|E(t)k\|^2$ is absolutely continuous on (a,b).

The main result on the representation of perturbation determinants is the following result of Krein [1]:

<u>Theorem 2.1</u>. Let A be a self-adjoint operator and let D be a self-adjoint trace class operator with the expansion

$$D = \sum_{n=1}^{\nu(D)} \lambda_n(D)(\ ,\phi_n)\phi_n ,$$

where $\{\phi_n\}$ is an orthonormal system and $\lambda_1(D),\lambda_2(D),\ldots,$ is an enumeration of the non-zero eigenvalues of D counted according to multiplicity. Set $B = A+D$.

There is an integrable function η (called the phase shift of the perturbation $A \to A+D$) defined on $\mathbb{R}$ providing the representation

$$\Delta_{B/A}(z) = \exp\int\frac{\eta(t)}{t-z}\,dt\ ,\ \operatorname{Im} z \neq 0\ . \qquad (2.9)$$

The function η has the following properties:

(i) $\operatorname{tr} D = \int_{-\infty}^{\infty}\eta(t)dt,$ (2.10)

(ii) $\int_{-\infty}^{\infty}|\eta(t)|dt \leq \sum_{n=1}^{\nu(D)}|\lambda_n(D)|$, and (2.11)

(iii) the function η decomposes into a sum

$$\eta = \sum_{n=1}^{\nu(D)} \eta_n\ , \qquad (2.12)$$

with

$$0 \le \eta_n \le 1, \quad \lambda_n(D) > 0$$
$$-1 \le \eta_n \le 0, \quad \lambda_n(D) < 0 \tag{2.13}$$

and

$$\int \eta_n(t)\,dt = \lambda_n(D) \; . \tag{2.14}$$

Proof. Set $A_0 = A$, $A_n = A_{n-1} + \lambda_n(D)(\;,\phi_n)\phi_n$, $(n = 1,2,\ldots)$. As we saw above, there is, for each n, the exponential representation

$$\Delta_{A_n/A_{n-1}}(z) = 1 + \lambda_n(D)((A_{n-1}-z)^{-1}\phi_n,\phi_n)$$
$$= \exp \int \frac{\eta_n(t)}{t-z}\,dt \; , \quad \operatorname{Im} z \neq 0, \tag{2.15}$$

where $0 \le \eta_n \le 1$, when $\lambda_n(D) > 0$, and $-1 \le \eta_n \le 0$, when $\lambda_n(D) < 0$. Equating first moments at infinity in (2.15) we obtain (2.14). Set $\eta = \sum_{n=1}^{\nu(D)} \eta_n$. The function η clearly has the properties (2.10)-(2.14).

From (2.2), for $\operatorname{Im} z \neq 0$ (fixed), $n = 1,2,\ldots$, there holds

$$\Delta_{B/A}(z) = \Delta_{B/A_n}\Delta_{A_n/A_{n-1}}\cdots\Delta_{A_1/A_0}(z)$$
$$= \det[I + D_n'(A_n-z)^{-1}] \exp\left\{\int \frac{\sum_{j=1}^{n}\eta_j(t)}{t-z}\,dt\right\} , \tag{2.16}$$

where we have set

$$D_n' = \sum_{j=n+1}^{\nu(D)} \lambda_j(D)(\;,\phi_j)\phi_j \; .$$

As $n \to \infty$,

$$\|D_n'(A_n-z)^{-1}\|_1 \to 0 \text{ and } \sum_{j=1}^{n} \eta_j \xrightarrow{L^1} \eta \; .$$

The representation (2.9) follows by taking the limit in equation (2.16) as $n \to \infty$. This completes the proof.

We mention some further properties of the phase shift.

6°. The function η appearing in the representation (2.9) is supported in the interval

$$[-\|A\| - \sum_{n=1}^{\nu(D)} |\lambda_n(D)| \ , \ \|A\| + \sum_{n=1}^{\nu(D)} |\lambda_n(D)|]$$

7°. The function η satisfies

$$|\eta(t)| \leq \nu(D) \ .$$

urther, if D is non-negative with $\nu(D)$ non-zero eigenvalues, then $\leq \eta \leq \nu(D)$.

8°. If the operator A restricted to the interaction space

$$H(A,D) = \bigvee\{A^j d : d \in R(D),\ j = 0,1,\dots\}$$

s purely singular, then η is integer valued. (Unlike the case of a ne-dimensional perturbation, the fact that η is integer valued does ot imply the operator A restricted to H(A,D) is purely singular.)

We conclude this section with a result of Krein [1] concerning he phase shift. This result will have an important two variable eneralization in the next chapter.

Proposition 2.1. Let A and D be self-adjoint operators on $\mathcal{H}$ with trace class. Let η denote the phase shift of the perturbation $\to$ A+D. Then for any polynomial p, we have

$$\mathrm{tr}[p(A+D) - p(A)] = \int_{-\infty}^{\infty} p'(t)\eta(t)\,dt \ . \qquad (2.17)$$

roof. It is obviously sufficient to establish (2.17) for the collec- ion of polynomials $p_j(t) = t^j$, $(j = 1,2,\dots)$. For $|\lambda|$ sufficiently nall, we can write

$$\Delta_{B/A}(\tfrac{1}{\lambda}) = \det([I - \lambda(A+D)][I-\lambda A]^{-1})$$

$$= \exp\ \mathrm{tr}(\log([I - \lambda(A+D)][I-\lambda A]^{-1}))$$

$$= \exp\ \mathrm{tr}(-\textstyle\int_0^\lambda [\frac{A+D}{I-\mu(A+D)} - \frac{A}{I-\mu A}]d\mu) .$$

The second equality follows from property 5° in Section 1. The last equality follows, for example, from Remark 9 of Gohberg and Krein [1, p. 163] .

A simple power series expansion leads to the identity

$$\Delta_{B/A}(\tfrac{1}{\lambda}) = \exp\left\{-\mathrm{tr}\left[\sum_{j=0}^{\infty} \frac{\lambda^{j+1}}{j+1} ((A+D)^{j+1} - A^{j+1})\right]\right\} . \qquad (2.18)$$

Elementary estimates involving the binomial expansion show

$$\|(A+D)^{j+1} - A^{j+1}\|_1 \leq (j+1)\|D\|_1 [\|A\|+\|D\|]^j .$$

Therefore, the series appearing in (2.18) will (for sufficiently small $|\lambda|$) converge in the trace norm $\| \ \|_1$. As a consequence

$$\Delta_{B/A}(\tfrac{1}{\lambda}) = \exp\left\{-\sum_{j=0}^{\infty} \frac{\lambda^{j+1}}{j+1} \mathrm{tr}[(A+D)^{j+1} - A^{j+1}]\right\} .$$

From (2.9), we obtain

$$\sum_{j=0}^{\infty} \frac{\lambda^{j+1}}{j+1} \mathrm{tr}((A+D)^{j+1} - A^{j+1}) = \sum_{j=0}^{\infty} \lambda^{j+1} \int_{-\infty}^{\infty} \eta(t) t^j dt .$$

This implies the desired result for the polynomials $\{p_j\}_{j=0}^{\infty}$ and completes the proof.

Notes

Section 1. In addition to the extensive work of Gohberg and Krein [1], the reader can find many of the properties of infinite determinants developed in the paper of Kuroda [1].

Section 2. The main reference for the perturbation determinant and the phase shift is the paper of Krein [1]. The paper of Aronszajn and Donoghue [1] presents a detailed description of the connection between the phase shift η and the measure $d\mu(t) = \lambda d\|E(t)k\|^2$ (in the case of one dimensional perturbations). The paper of Donoghue [1] sheds further light on the subject of one-dimensional perturbations. In [1], Carey has generalized the phase shift to an operator valued function. The result in Proposition 2.1 extends beyond the case of polynomials to sufficiently smooth functions (See, Krein [1]).

CHAPTER 5

THE PRINCIPAL FUNCTION

In this chapter we will study an invariant associated with every operator that has a trace class self-commutator. This invariant is a real valued integrable function on the plane which is referred to as the principal function of the operator. The principal function was first introduced by Pincus [1] in connection with the diagonalization of certain self-adjoint singular integral operators. In their paper [1], Helton and Howe approached the study of operators with a trace class self-commutator by representing an associated tracial bilinear form as integration up against a signed measure in the plane. This measure was shown by Pincus [3] (See also Carey and Pincus [3]) to have derivative equal to the principal function. A rich theory based on the work of these authors has evolved.

The development below is limited in scope. We have elected to describe a "computable" method of obtaining the principal function for seminormal operators with a trace class self-commutator. This description is based on the phase shift that was discussed in the preceding chapter. The final section of the chapter contains some recent results of Berger [1] relating the size of the principal function to the existence of cyclic vectors.

1. Tracial Bilinear Forms.

A pair A,B of operators on a Hilbert space $\mathcal{H}$ is called an almost commuting pair (a.c. pair) in case the commutator $[A,B] = AB-BA$ is a trace class operator. In case T is an operator on $\mathcal{H}$ such that T^*,T is an a.c. pair, then the operator T is said to be nearly normal. If $T = X + iY$ is the Cartesian form of the operator T, then T is nearly normal if and only if X,Y is a self-adjoint a.c. pair.

Let A,B be an a.c. pair. The notation $\mathcal{P}(A,B)$ will denote the lgebra (non-closed) generated by the operators A,B and the identity. he elements in $\mathcal{P}(A,B)$ are formally polynomials in the operators A and . If R,S belong to the algebra $\mathcal{P}(A,B)$, then the commutator [R,S] lies n the trace class. The <u>tracial bilinear form</u> associated with the .c. pair A,B is the bilinear form defined for R,S in $\mathcal{P}(A,B)$ by

$$(R,S) = \operatorname{tr} i[R,S] \quad . \tag{1.1}$$

he factor i is built into the definition (1.1) mainly for convenience.

The notation $\mathcal{P}(\mathbb{R}^2)$ will be used for the algebra of polynomials ver $\mathbb{C}$ in the two commuting indeterminants x,y. A typical element in $(\mathbb{R}^2)$ can be written as a finite sum of the form

$$p(x,y) = \Sigma a_{ij} x^i y^j \ , \tag{1.2}$$

ere a_{ij} are complex numbers. Let X,Y be a self-adjoint a.c. pair. e associate with the element p in $\mathcal{P}(\mathbb{R}^2)$, which has the form (1.2), ne element in $\mathcal{P}(X,Y)$ defined by

$$p(X,Y) = \sum_{i,j} a_{ij} X^i Y^j \quad . \tag{1.3}$$

ith this convention, the tracial bilinear form associated with the .c. pair X,Y, induces the bilinear form

$$(p,q) = \operatorname{tr} i[p(X,Y), q(X,Y)] \tag{1.4}$$

 elements p,q in $\mathcal{P}(\mathbb{R}^2)$.

The rule for associating p(X,Y) with the element $p \in \mathcal{P}(\mathbb{R}^2)$ is t unique. Any rule which substitutes X,Y for x,y, respectively, ads to an operator that differs from p(X,Y), given as in (1.3), by trace class operator. As a result these rules induce the same bi- near form (1.4) on $\mathcal{P}(\mathbb{R}^2)$.

The most important property possessed by the bilinear form (1.4) $\mathcal{P}(\mathbb{R}^2)$ is the <u>collapsing property</u>. This means the following: If p and q are polynomials of a single variable (we write this $p,q \in \mathcal{P}(\mathbb{R})$) and $r \in \mathcal{P}(\mathbb{R}^2)$, then

$$(p\circ r,\ q\circ r) = \operatorname{tr} i[p(r(X,Y)),\ q(r(X,Y))] = 0 .$$

A representation of collapsing bilinear forms on $\mathcal{P}(\mathbb{R}^2)$ provides the key to the representation of the tracial bilinear form. This result on collapsing bilinear forms is provided in the following result of Helton, Howe and Wallach. The following proof (due to N. Wallach) appears in Helton and Howe [1]. We include the argument for the sake of completeness.

Theorem 1.1. Let $(\ ,\ )$ be a collapsing form on the algebra $\mathcal{P}(\mathbb{R}^2)$. Then there is a unique linear functional ℓ on $\mathcal{P}(\mathbb{R}^2)$ such that

$$(p,q) = \ell(J(p,q)), \tag{1.5}$$

where $J(p,q) = \frac{\partial p}{\partial x}\frac{\partial q}{\partial y} - \frac{\partial q}{\partial x}\frac{\partial p}{\partial y}$ denotes the Jacobian of the transformation $(x,y) \to (p(x,y),\ q(x,y))$ on $\mathbb{R}^2$.

Proof. The uniqueness of the representation (1.5) is obvious. For $q \in \mathcal{P}(\mathbb{R}^2)$ define

$$\ell(q) = (x,Q) ,$$

where Q is any polynomial satisfying $\frac{\partial Q}{\partial y} = q$. Note that ℓ is well defined. In fact, if $\frac{\partial Q_1}{\partial y} = \frac{\partial Q_2}{\partial y} = q$, then $Q_1 - Q_2$ depends only on x. By the collapsing property, $(x,Q_1-Q_2) = 0$, therefore, $(x,Q_1) = (x,Q_2)$.

It suffices to show the collapsing form

$$(\ ,\ )_1 = (\ ,\ ) - \ell \circ J$$

is zero. Note $(x,q)_1 = 0$, for all q in $\mathcal{P}(\mathbb{R}^2)$.

First it is shown that, if r,s satisfy

$$(r,q)_1 = (s,q)_1 = 0, \text{ for all } q \text{ in } \mathcal{P}(\mathbb{R}^2),$$

then

$$(rs,q)_1 = 0, \text{ for all } q \text{ in } \mathcal{P}(\mathbb{R}^2).$$

Let $\alpha,\beta \in \mathbb{C}$ and $q \in \mathcal{P}(\mathbb{R}^2)$. Using the collapsing property

$$((\alpha r+\beta s+q)^2, \ (\alpha r+\beta s+q))_1 = 0 .$$

Consequently,

$$\alpha^2(r^2,q)_1 + \beta^2(s^2,q)_1 + 2\alpha\beta(rs,q)_1 + 2\alpha(rq,q)_1 + 2\beta(sq,q)_1 = 0 .$$

Taking the coefficient of $\alpha\beta$, we obtain $(rs,q)_1 = 0$.

Next note, if $\alpha \in \mathbb{C}$ and $n \geq 1$,

$$0 = ((y+\alpha x), \ (y+\alpha x)^n)_1 = \sum_{j=0}^{n} \binom{n}{j} \alpha^j (y, x^j y^{n-j})_1 .$$

This implies, $(y,x^j y^{n-j})_1 = 0$, $j = 1,\ldots,n; n = 1,2,\ldots$ and , therefore, $(y,q)_1 = 0$, for all $q \in \mathcal{P}(\mathbb{R}^2)$. Since $(x,q)_1 = (y,q)_1 = 0$, for all $q \in \mathcal{P}(\mathbb{R}^2)$, and products of elements r satisfying $(r,q)_1 = 0$, for all $q \in \mathcal{P}(\mathbb{R}^2)$, also have this property, then $(\ , \ )_1 = 0$. This completes the proof.

In their paper [1], Helton and Howe showed that the linear functional ℓ which provides the representation (1.5) for a tracial bilinear form is obtained by integrating the Jacobian with respect to a measure. Specifically, these authors establish the following:

<u>Theorem 1.2</u>. Let X,Y be an a.c. pair of self-adjoint operators on the Hilbert space $\mathcal{H}$. There exists a regular signed Borel measure P with compact support on $\mathbb{R}^2$ such that for $p,q \in \mathcal{P}(\mathbb{R}^2)$,

$$(p,q) = \operatorname{tr} i[p(X,Y), \ q(X,Y)] = \iint_{\mathbb{R}^2} J(p,q)\,dP. \quad (1.6)$$

We will not present the details of the proof of Theorem 1.2 in these notes. The construction of the measure P is carefully described in an appendix to Helton and Howe's paper [1].

Immediately after Helton and Howe obtained the respresentation (1.6), Pincus [3] established that the measure P is absolutely continuous with respect to planar Lebesgue measure. This was basically accomplished by noting the connection between the tracial bilinear form and the determining (determinant) function approach to almost commuting pairs which Pincus had been vigorously pursuing for several years. The "derivative" g of the measure P satisfying

$$dP = \frac{1}{2\pi} g(x,y)dxdy$$

is called the principal function of the almost commuting pair.

The reader will recognize the representation

$$\operatorname{tr} i[p(X,Y),q(X,Y)] = \frac{1}{2\pi} \iint_{\mathbb{R}^2} J(p,q)g(x,y)dxdy$$

of the tracial bilinear form as the two variable analogue of the representation of the linear functional

$$\operatorname{tr}[p(A+D) - p(A)] = \int_{\mathbb{R}} p'(t)\eta(t)dt, \quad p \in \mathcal{P}(\mathbb{R}) \quad .$$

This latter representation, involving the phase shift of the trace class perturbation $A \to A+D$, was obtained in Proposition 2.1 of the preceding chapter. As a consequence, it is natural to view the principal function as the two variable phase shift. This connection will be even more apparent after the construction of the principal function for a nearly, normal seminormal operator in the next section.

2. The Principal Function for Hyponormal Operators.

Let $H = X + iY$ be a nearly, normal hyponormal operator defined on a Hilbert space $\mathcal{H}$. In this section we will construct the principal function $g = g_H$ of the operator H. In other words, g will be an integrable function on $\mathbb{R}^2$ such that for $p,q \in \mathcal{P}(\mathbb{R}^2)$

$$(p,q) = \operatorname{tr} i[p(X,Y), q(X,Y)] = \frac{1}{2\pi} \iint_{\mathbb{R}^2} J(p,q)g(x,y)\,dx\,dy.$$

The construction of the principal function g outlined below will depend on the singular integral representation of H as described in Chapter 2. In restricting to the hyponormal case, we have several advantages. One such advantage is that the existence of the symbol homomorphisms is trivial. At the same time, when the operator is given its singular integral representation, the method leads to a fairly precise description of the principal function. The method we follow is basically the method of Carey and Pincus [3] specialized to the case of a hyponormal pair.

For the remainder of this section we will assume that the hyponormal operator $H = X + iY$ is represented as a singular integral operator on the space $\mathcal{H}_0$ as described in Theorem 3.1 of Chapter 2. Certain relevant properties of this representation are now recalled.

The space $\mathcal{H}_0 = \mathcal{H}_1 \oplus \mathcal{H}_2$ is the direct sum of the direct integral spaces

$$\mathcal{H}_1 = \int_E \oplus \mathcal{H}_1(t)dt; \quad \mathcal{H}_2 = \int \oplus \mathcal{H}_2(t)d\nu .$$

We recall that the spaces $\mathcal{H}_1(t)$ $(t \in E)$ are closed subspaces of the space $\mathcal{K}$. For definiteness, we assume the spaces $\mathcal{H}_2(t)$ $(t \in E)$ are subspaces of the fixed Hilbert space $\mathcal{K}_\infty$. The operator X becomes the diagonal operator

$$X = \begin{bmatrix} \Lambda_1 & 0 \\ 0 & \Lambda_2 \end{bmatrix} ,$$

where Λ_i are the decomposable operators $\Lambda_i f(t) = tf(t)$ defined for $f \in \mathcal{H}_i$ $(i=1,2)$. The self-commutator $D = [H^*,H]$ has the form

$D = \begin{bmatrix} 2Z & 0 \\ 0 & 0 \end{bmatrix}$, where Z is the integral operator

$$Zf(x) = \frac{B_1(x)P_1(x)}{2\pi} \int_E B_1(t)f(t)dt .$$

This representation of D is presented in Corollary 3.1 of Chapter 2. It will be important to realize that the decomposable operator $B_1 = \int_E \oplus B_1^2(t)dt$ has the property that $B_1^2(t)$ is trace class almost everywhere (see, Corollary 4.1 of Chapter 2).

Finally, we recall the form of the symbols $S_X^{\pm}$ in this singular integral representation of H. For any operator A in $\mathcal{C}^*(H)$, the symbols $S_X^{\pm}(A)$ have the "decomposable" forms

$$S_X^{\pm}(A) = \begin{bmatrix} \int_E \oplus A_1^{\pm}(t)dt & A_{12}^{\pm} \\ A_{21}^{\pm} & \int \oplus A_2^{\pm}(t)d\nu \end{bmatrix} , \quad (2.1$$

where $A_{12}^{\pm} : \mathcal{H}_2 \to \mathcal{H}_1$, $A_{21}^{\pm} : \mathcal{H}_1 \to \mathcal{H}_2$ are bounded operators satisfying $A_{12}^{\pm}\Lambda_2 = \Lambda_1 A_{12}^{\pm}$ and $\Lambda_2 A_{21}^{\pm} = A_{21}^{\pm}\Lambda_1$. These latter intertwining identities mean the following: the operators $A_{12}^{\pm} : \mathcal{H}_2 \to \mathcal{H}_1$ have the forms

$$A_{12}^{\pm}f(t) = A_{12}^{\pm}(t)f(t), \; t \in E,$$

where $A_{12}^{\pm}(t) : \mathcal{H}_2(t) \to \mathcal{H}_1(t)$ and $(A_{12}^{\pm}(t)f,g)$ is Borel measurable for $g \in \mathcal{R}$, $f \in \mathcal{K}_\infty$. Moreover, the norms of $A_{12}^{\pm}(t)$ $(t \in E)$ are bounded. Similarly, the operators $A_{21}^{\pm} : \mathcal{H}_1 \to \mathcal{H}_2$ have representations

$$A_{21}^{\pm}f(t) = A_{21}^{\pm}(t)f(t), \; t \in \mathbb{R} .$$

We will use the notation

$$A^{\pm}(t) = \begin{bmatrix} A_1^{\pm}(t) & A_{12}^{\pm}(t) \\ A_{21}^{\pm}(t) & A_2^{\pm}(t) \end{bmatrix} , \quad (t \in E) .$$

Note for $t \in E$, the operators $A^{\pm}(t)$ act on $\mathcal{H}_1(t) \oplus \mathcal{H}_2(t)$.

In Corollary 3.2 of Chapter 2, it was shown that

$$S_X^+(Y) - S_X^-(Y) = \begin{bmatrix} \int_E^{\oplus} B_1^2(t)\,dt & 0 \\ 0 & 0 \end{bmatrix}$$

The definition of the principal function will be made in terms of the phase shifts of the perturbations $Y^-(t) \to Y^+(t) = Y^-(t) + B(t)\,(t \in E)$, where we have set

$$B(t) = \begin{bmatrix} B_1(t) & 0 \\ 0 & 0 \end{bmatrix} .$$

To this end, we can assume that E is a Borel set and that $x \to Y^-(x)$, $x \to B^2(x)$ are weakly Borel measurable. Moreover, it can be assumed that for every x in E, the perturbation

$$Y^-(x) \to Y^-(x) + B^2(x) = Y^+(x) \tag{2.2}$$

is trace class.

For $x \in E$, let $\tilde{g}(x,\ ) \in L^1(\mathbb{R})$ be the phase shift of the pertubation (2.2). Accordingly, $\tilde{g}(x,\ )$ satisfies

$$\det(I + B^2(x)[Y^-(x)-z]^{-1}) = \exp \int \frac{\tilde{g}(x,y)}{y-z}\,dy, \ \operatorname{Im} z \neq 0 . \tag{2.3}$$

It should be emphasized that the function $\tilde{g}(x,y)$ defined on $E \times \mathbb{R}$ is not a priori planar Lebesgue measurable. It requires a little work to select a measurable principal function. This is the content of the following:

<u>Lemma 2.1</u>. There is a integrable Borel measurable function g defined on $E \times \mathbb{R}$ such that for every $x \in E$,

$$g(x,\ ) = \tilde{g}(x,\ ) . \tag{2.4}$$

The equality in (2.4) is intended as equality in $L^1(\mathbb{R})$.

Proof. The spaces $\mathcal{N}_1(t)$ $(t \in E)$ are subspaces of the space $\mathcal{R}$ which was defined in Chapter 2 as the closure of the range of D. Let $\{\phi_n\}_{n=1}^{\nu(D)}$ be an orthonormal basis in $\mathcal{R}$ and let P_n $(n = 1,2,\ldots,\nu(D))$ be the orthogonal projection of $\mathcal{R} \oplus \mathcal{H}_\infty$ onto the space spanned by $\phi_n \oplus 0$.

For $x \in E$, we set

$$Y_0(x) = Y^-(x), \quad Y_n(x) = Y_{n-1}(x) + P_n B(x) P_n, \quad n = 1,2,\ldots \quad .$$

For $(x,y,\tau) \in E \times \mathbb{R} \times \mathbb{R}^+$ and $n = 1,2,\ldots,$ define

$$g_n(x,y,\tau) = \frac{1}{\pi} \operatorname{Im}\{\log \det(I + P_n B(x) P_n [Y_{n-1}(x) - (y+i\tau)]^{-1})\} \; .$$

The functions g_n are continuous in the variables y,τ and Borel measurable in the variable x. Consequently, g_n is Borel measurable on $E \times \mathbb{R} \times \mathbb{R}^+$ (see, Rudin [1, Chapter 7, Exercise 8]).

Let F_n be the set in $E \times \mathbb{R}$ on which

$$g_n(x,y) \equiv \lim_{\tau \downarrow 0} g_n(x,y,\tau)$$

exists. Note for x fixed in E, $g_n(x,y)$ exists for almost every $y \in \mathbb{R}$. In fact, $g_n(x,\ )$ is the phase shift of the perturbation $Y_{n-1}(x) \to Y_n(x)$. The function g_n is Borel measurable on the Borel set F_n.

On the set $F = \bigcap_{n=1}^{\nu(D)} F_n$, we define the non-negative Borel function

$$g(x,y) = \sum_{n=1}^{\nu(D)} g_n(x,y) \quad .$$

Again we remark, for x fixed in E, the function $g(x,y)$ is defined at almost every y in $\mathbb{R}$. In fact, equation (2.3) holds with $\tilde{g}(x,\ )$ replaced by $g(x,\ )$. This implies (2.4).

From Fubini's Theorem, equation (2.4) and equation (2.10) of Chapter 4,

$$\iint_{E\times\mathbb{R}} g(x,y)\,dx\,dy = \int_E \int_{\mathbb{R}} \tilde{g}(x,y)\,dy\,dx$$

$$= \int_E \operatorname{tr} B_1^2(x)\,dx < \infty \quad .$$

This establishes the integrability of g and completes the proof.

In the sequel, it will be assumed that the principal function g of the hyponormal operator $H = X + iY$ is defined to be zero off $E \times \mathbb{R}$. When speaking of g we will be referring to the equivalence class of g in $L^1(\mathbb{R}^2)$.

We now turn to the process of showing that the principal function meets the requirements discussed in the introductory section.

Proposition 2.1. Let $H = X + iY$ be a hyponormal operator with trace class self-commutator. Assume H has the singular integral representation as described above and let g be the principal function of the operator H. Then, for $n \geq 1$,

$$\begin{aligned}\operatorname{tr} i[X^nY - YX^n] &= \frac{1}{2\pi} n\iint x^{n-1} g(x,y)\,dxdy \\ &= \frac{1}{2\pi} n\int_E x^{n-1} \operatorname{tr}[Y^+(x) - Y^-(x)]dx\ .\end{aligned}$$

Proof. Let $n \geq 1$. It is easy to verify

$$\operatorname{tr} i[X^nY - YX^n] = \operatorname{tr}(nX^{n-1} i[X,Y])\ .$$

Moreover, the operator $nX^{n-1}i[X,Y]$ on the Hilbert space $\mathcal{H}_0$ has the form $\begin{bmatrix} Z_n & 0 \\ 0 & 0 \end{bmatrix}$, where

$$Z_n f(x) = \frac{1}{2\pi} nx^{n-1} B_1(x) P_1(x) \int_E B_1(t) f(t)\,dt\ .$$

Consequently,

$$\begin{aligned}\operatorname{tr}(nX^{n-1}i[X,Y]) &= \frac{1}{2\pi} n\int_E x^{n-1} \operatorname{tr} B_1^2(x)\,dx \\ &= \frac{1}{2\pi} n\int_E x^{n-1} \operatorname{tr}[Y^+(x) - Y^-(x)]dx \\ &= \frac{1}{2\pi} n\int_E x^{n-1} (\int g(x,y)\,dy)\,dx\end{aligned}$$

This completes the proof.

Proposition 2.2. Assume the hypothesis of Proposition 2.1. Then, for m,n = 1,2,... ,

$$\operatorname{tr} i[X^n Y^m - Y^m X^n] = \frac{1}{2\pi} nm \iint x^{n-1} y^{m-1} g(x,y)\,dx\,dy .$$

Proof. For $s > \|Y\|$, we observe

$$i[(-X)Y_s^{-1} - Y_s^{-1}(-X)] = Y_s^{-1} i[X,Y] Y_s^{-1} ,$$

where as usual $Y_s = Y-s$.

Applying Proposition 2.1 (with fixed $n \geq 1$) to the hyponormal operator $H(s) = -X + iY_s^{-1}$, we obtain

$$\operatorname{tr} i[X^n Y_s^{-1} - Y_s^{-1} X^n] = \frac{1}{2\pi} n \int_E x^{n-1} \operatorname{tr}([Y^+(x)-s]^{-1} - [Y^-(x)-s]^{-1})\,dx .$$

Replacing Y_s^{-1} in this last identity with the series

$$(Y-s)^{-1} = - \sum_{m=0}^{\infty} \frac{Y^m}{s^{m+1}} ,$$

we have

$$\sum_{m=0}^{\infty} \frac{\operatorname{tr} i[X^n Y^m - Y^m X^n]}{s^{m+1}} = \sum_{m=0}^{\infty} \frac{1}{s^{m+1}} \frac{1}{2\pi} \int_E n x^{n-1} \operatorname{tr}[(Y^+)^m(x) - (Y^-)^m(x)]\,dx .$$

(The fact that the trace and integration may be interchanged with the summation is not difficult to verify.) Equating coefficients of the two series, we obtain

$$\operatorname{tr} i[X^n Y^m - Y^m X^n] = \frac{1}{2\pi} n \int_E x^{n-1} \operatorname{tr}[(Y^+)^m(x) - (Y^-)^m(x)]\,dx . \quad (2.5)$$

From Proposition 2.1 of Chapter 4,

$$\operatorname{tr}[(Y^+)^m(x) - (Y^-)^m(x)] = m \int y^{m-1} g(x,y)\,dy .$$

This last identity may be substituted into equation (2.5) to complete the proof.

At this stage, we have basically established

$$\operatorname{tr} i[p(X),q(Y)] = \frac{1}{2\pi}\iint \frac{\partial p}{\partial x}\frac{\partial q}{\partial y}\, g(x,y)\,dxdy \; ,$$

when p and q are polynomials of a single variable. The key to passing to the representation of the tracial bilinear form for elements in $\mathcal{P}(\mathbb{R}^2)$ is Theorem 1.1.

The main result in this section is the following:

<u>Theorem 2.1</u>. Let $H = X + iY$ be a hyponormal operator with trace class self-commutator and g its principal function. Then for $p,q \in \mathcal{P}(\mathbb{R}^2)$

$$(p,q) = \operatorname{tr} i[p(X,Y),q(X,Y)] = \frac{1}{2\pi}\iint J(p,q)g(x,y)\,dxdy \; . \quad (2.9)$$

Proof. By Theorem 1.1, $(p,q) = \ell(J(p,q))$ for some linear functional ℓ on $\mathcal{P}(\mathbb{R}^2)$. After Proposition 2.2,

$$\begin{aligned}(x^n,y^m) &= \frac{1}{2\pi}\iint J(x^n,y^m)g(x,y)\,dxdy\\ &= \frac{1}{2\pi}\, nm \iint x^{n-1}y^{m-1}g(x,y)\,dxdy\\ &= nm\; \ell(x^{n-1}y^{m-1}) \quad .\end{aligned}$$

Thus ℓ agrees with the linear functional defined on $p \in \mathcal{P}(\mathbb{R}^2)$ by

$$\ell(p) = \frac{1}{2\pi}\iint p(x,y)g(x,y)\,dxdy \quad .$$

This completes the proof.

. <u>Properties of the Principal Function.</u>

In this section we record some of the basic properties of the principal function associated with a hyponormal operator. Throughout this section, it will be assumed that $H = X + iY$ is a nearly normal, hyponormal operator and g its principal function constructed as in the preceding section.

1°. Let $D = [H^*,H]$ and let n_1 denote the spectral multiplicity function of the operator X restricted to the space

$$\mathcal{H}_1 = \bigvee\{X^n d : d \in R(D),\ n = 0,1,2,\dots\} .$$

Then

$$0 \leq g(x,\ ) \leq n_1(x) . \tag{3.1}$$

We remark that the operator X restricted to $\mathcal{H}_1$ is absolutely continuous and (3.1) holds a.e. with respect to Lebesgue measure on $\mathbb{R}$. The estimate (3.1) follows from the obvious inequality

$$\operatorname{rank}(B_1^2(x)) \leq n_1(x)$$

plus the fact that the phase shift is dominated by the rank of the pertubation (see Remark 7° in Section 2 of the preceding chapter).

Also we note, if the multiplicity function n of the operator X is finite, then g is integer valued. Indeed, in this case, both operator $Y^{\pm}(x)$ act on a finite dimensional space.

2°. The principal function satisfies

$$0 \leq g(x,y) \leq \operatorname{rank}[H^*,H] . \tag{3.2}$$

The estimate (3.2) follows immediately from (3.1) and the inequality $n_1(x) \leq \operatorname{rank}\ [H^*,H]$.

3°. Let E (respectively F) be a support set for the self-adjoint operator X_{ac} (respectively Y_{ac}). The principal function g is supported on $E \times F$. This remark is seen as follows: The construction of g shows it is supported on $E \times \mathbb{R}$. On the other hand we could have constructed the principal function by diagonalizing the real part of the hyponormal operator $H' = -iH = Y - iX$. Consequently, g is supported on $\mathbb{R} \times F$.

4°. Let β be a Borel set in $\mathbb{R}$. Assume $\int \lambda dE(\lambda)$ is the spectral resolution of X. Let H_β denote the hyponormal operator H cut down to he space $\mathcal{H}_\beta = E(\beta)\mathcal{H}$. Of course, H_β is a hyponormal operator with race class self-commutator. Denote by g_β the principal function of he operator H_β . Then

$$g_\beta = \chi_{\beta\times\mathbb{R}}\, g, \tag{3.3}$$

here $\chi_{\beta\times\mathbb{R}}$ is the characteristic function of $\beta \times \mathbb{R}$. The identity 3.3) is immediate from the definitions of g and g_β given as in emma 2.1.

5°. Assume H is pure. We denote by supp(g) the (essential) upport of g in $\mathbb{C}$ (identified with $\mathbb{R}^2$) . Thus a complex number z is n supp(g) if and only if every neighborhood of z intersects the set x+iy$|$g(x,y) $\neq$ 0} in a set of positive measure. We have the identity

$$\sigma(H) = \text{supp}(g) \ . \tag{3.4}$$

ne equality (3.4) is proved as follows: Suppose $z = x + iy \in \sigma(H)$ nd g vanishes on the square $\Delta \times \delta = [a,b] \times [c,d]$ which has center x,y). Let $X = \int\lambda dE(\lambda)$ and $Y_\Delta = \int\lambda dF^\Delta(\lambda)$ be the spectral resolutions X and $Y_\Delta = E(\Delta)YE(\Delta)$ on $\mathcal{H}$ and $\mathcal{H}_\Delta = E(\Delta)\mathcal{H}$, respectively. According 4°, the principal function $g_{\Delta,\delta}$ of the operator $H_{\Delta,\delta} = F^\Delta(\delta)H_\Delta F^\Delta(\delta)$ the space $\mathcal{H}_{\Delta,\delta} = F^\Delta(\delta)E(\Delta)\mathcal{H}$ is zero. However, by Lemma 1.1 of apter 3, the operator $H_{\Delta,\delta}$ is a pure hyponormal operator and from rollary 1.2 of Chapter 3, the space $\mathcal{H}_{\Delta,\delta}$ is non-trivial. This is a ntradiction. We conclude $\sigma(H) \subset \text{supp}(g)$. A similar argument involv- g cut downs establishes the reverse inclusion $\text{supp}(g) \subset \sigma(H)$.

The relation between the principal function g and the essential ectrum is not fully understood. In the case where $[H^*,H]$ has nk one, then this relationship has been determined (see, e.g. Carey d Pincus [2]).

It is also known, that on a component of the essential resolvent, e Fredholm index satisfies

$$\text{index}(H - (x + iy)) = -g(x,y) \ . \tag{3.5}$$

The identity (3.5) is established in Helton and Howe [1].

We conclude this section with a change of variables formula for the principal function.

6°. Let α,β be real valued polynomials in $\mathcal{P}(\mathbb{R}^2)$. Let A,B be the a.c. pair of operators in $\mathcal{C}^*(H)$ defined by

$$A = \alpha(X,Y), \quad B = \beta(X,Y) \ .$$

Note that A and B may fail to be self-adjoint; however, they are equal to self-adjoint operators plus elements from the trace ideal.

We introduce the tracial bilinear form defined for elements $p,q \in \mathcal{P}(\mathbb{R}^2)$ by

$$(p,q)_{\alpha,\beta} = \text{tr}\ i[p(A,B),q(A,B)] \ .$$

This tracial bilinear form has the representation

$$(p,q)_{\alpha,\beta} = \frac{1}{2\pi} \iint_{\mathbb{R}^2} J(p,q)g_{\alpha,\beta}(x,y)dxdy \ ,$$

where

$$g_{\alpha,\beta}(x,y) = \sum_{\substack{\alpha(\xi,\eta) = x \\ \beta(\xi,\eta) = y}} \text{sign}(J(\alpha,\beta))g(\xi,\eta). \tag{3.6}$$

Here sign$(J(\alpha,\beta))$ denotes the sign of the Jacobian of the mapping τ defined on $\mathbb{R}^2$ by

$$\tau(\xi,\eta) = (\alpha(\xi,\eta), \ \beta(\xi,\eta)) .$$

The sign is taken to be zero when $J(\alpha,\beta) = 0$.

The equality in (3.6) follows from the identities

$$
\begin{aligned}
(p,q)_{\alpha,\beta} &= (p\circ\tau,\ q\circ\tau) \\
&= \frac{1}{2\pi}\iint_{\mathbb{R}^2} J(p\circ\tau, q\circ\tau)\ g(x,y)\,dxdy \\
&= \frac{1}{2\pi}\iint_{\mathbb{R}^2} J(p,q)(\alpha(\xi,\eta),\beta(\xi,\eta))\ J(\alpha,\beta)\ g(\xi,\eta)\,d\xi d\eta \\
&= \frac{1}{2\pi}\iint_{\mathbb{R}^2} J(p,q)\ g_{\alpha,\beta}(x,y)\,dxdy.
\end{aligned}
$$

ne case of particular interest is the following: Let f be a non-onstant complex polynomial,

$$f(z) = \sum_{k=0}^{N} a_k z^k .$$

onsider the nearly normal operator

$$f(H) = \sum_{k=0}^{N} a_k H^k = U + iV, \tag{3.7}$$

here U,V are the real and imaginary parts of $f(H)$.

The above change of variables result leads to the representation

$$\operatorname{tr} i[p(U,V),q(U,V)] = \frac{1}{2\pi}\iint_{\mathbb{R}^2} J(p,q) g_{f(H)}(x,y)\,dxdy,\ p,q \in \mathcal{P}(\mathbb{R}^2) ,$$

iere

$$g_{f(H)}(x,y) = \sum_{f(\xi+i\eta)=x+iy} g(\xi,\eta). \tag{3.8}$$

Explicit Descriptions of the Principal Function.

In this section we will describe how the principal function can computed for certain hyponormal operators which are given as singu-r integral operators. After the remark concerning the principal nction of a cut down (see 4° in Section 3), it is clearly sufficient describe the principal function of the hyponormal operator $= X + iY$ cut down to a set where the spectral multiplicity of X is iformly equal to a constant (possibly infinity). For the most part, we will concentrate on the case where this spectral multiplicity is finite.

Fix $n (1 \le n < \infty)$. Let $E \subset \mathbb{R}$ be a bounded measurable set. The notation $L^2(E:\mathbb{C}^n)$ will be used for the Lebesgue space of $\mathbb{C}^n$-valued square integrable functions on E. The notation $L^\infty(E:M_n)$ will be used for the space of n×n-matrix-valued essentially bounded functions on E. Each element $A \in L^\infty(E:M_n)$ induces the multiplication operator

$$Af(t) = A(t)f(t)$$

on elements $f \in L^2(E:\mathbb{C}^n)$. The reader will recognize that $L^2(E:\mathbb{C}^n)$ is a direct integral space and that $L^\infty(E:M_n)$ are the decomposable operators on $L^2(E:\mathbb{C}^n)$.

Let $A,B \in L^\infty(E:M_n)$ with $A(t) = A^*(t)$ and $B(t) \ge 0$ almost everywhere. Define the hyponormal singular integral operator S on $L^2(E:\mathbb{C}^n)$ by

$$Sf(x) = xf(x) + i\left[A(x)f(x) + \frac{B(x)}{\pi i} \int_E \frac{B(t)f(t)}{x-t}\,dt\right] . \qquad (4.1)$$

The operator S resembles the form of an arbitrary hyponormal operator $H = X + iY$, when H is cut down to a set where the spectral multiplicit of X is uniformly equal to n.

If we write (4.1) in the Cartesian form $S = X + iY$ and compare with the representation in Theorem 3.1 of Chapter 2, then we obtain

$$S_X^{\pm}(Y) = A \pm B^2 .$$

The definition of the principal function g for the operator S (see equation (2.3) and Lemma 2.1) implies, for almost every $x \in E$ and $\operatorname{Im} z \neq 0$,

$$\det([A(x) + B^2(x) - z][A(x) - B^2(x) - z]^{-1} = \exp\int \frac{g(x,y)}{y-z}\,dy . \quad (4.2)$$

Of course, the identity (4.2) just restates the fact that $g(x,\ )$ is, for almost every $x \in E$, the phase shift of the pertubation $A(x) - B^2(x) \to A(x) + B^2(x)$.

In the case n = 1, it is easy to see that the function g(x,) is the characteristic function of the interval $[A(x)-B^2(X), A(x)+B^2(x)]$. We summarize this case:

Proposition 4.1. Let E be a bounded measurable subset of $\mathbb{R}$. Let $a,b \in L^\infty(E)$ with a(t) real valued and $b(t) \neq 0$ almost everywhere. Let S denote the hyponormal singular integral operator defined on $L^2(E)$ by

$$Sf(x) = xf(x) + i\left[a(x)f(x) + \frac{b(x)}{\pi i}\int_E \frac{\overline{b(t)}f(t)}{x-t}\,dt\right].$$

The principal function of the operator S is the characteristic function of the set

$$\mathscr{J} = \{(x,y) \in \mathbb{R}^2 : x \in E,\ a(x) - |b(x)|^2 < y < a(x) + |b(x)|^2\} .$$

This proposition is a direct consequence of the above discussion. The operator S represented in the above proposition is (up to unitary equivalence) the most general pure hyponormal operator with a rank one self-commutator whose real part has uniform spectral multiplicity equal to one. Remark 5° of the preceding section can be used with Proposition 4.1 to determine the spectrum of the operator S.

Although, in some sense, (4.2) describes the principal function of the operator (4.1) it is not easy to describe the principal function in much detail. We saw in Section 2 of the preceding chapter, if the rank of $B^2(x)$ equals one, then a slightly more detailed description of the principal function is possible. In theory, we can realize the pertubation $A(x) - B^2(x) \to A(x) + B^2(x)$ as a sequence of at most n rank one non-negative pertubations; however, this does not afford any sweeping statements about g(x,).

In case A(x) commutes with $B^2(x)$, then these operators can be simultaneously diagonalized. Assuming this simultaneous diagonalization we let $a_1(x) \ldots a_n(x)$ and $b_1(x) \ldots b_n(x)$ be the eigenvalues of

A and B respectively. Then $g(x, \ )$ is the sum of the characteristic functions of the intervals

$$[a_i(x) - b_i^2(x), a_i(x) + b_i^2(x)], \quad i = 1,\dots,n.$$

This clearly implies the following:

<u>Proposition 4.2.</u> Let $B \in L^\infty(E:M_n)$ with $B \geq 0$. Let S be the hyponormal singular integral operator on $L^2(E:\mathbb{C}^n)$ defined by

$$Sf(x) = xf(x) + \frac{B(x)}{\pi} \int_E \frac{B(t)f(t)}{x-t} dt . \tag{4.3}$$

Assume $0 \leq b_1(x) \leq \dots \leq b_n(x)$ is an enumeration of the eigenvalues of $B(x)$. Then the principal function of the operator S is the function

$$g(x, \ ) = \sum_{i=1}^{n} \chi_{[-b_i^2(x), b_i^2(x)]} .$$

We remark that there is an analogue of Proposition 4.2 in the case $n = \infty$. In this case the operator S defined by (4.3) acts on the space $L^2(E:\ell_2)$. In order that the operator S have a trace class self-commutator, we must further assume

$$\int_E \operatorname{tr} B^2(x)dx < \infty .$$

5. An Estimate on the Principal Function.

In Section 1 of this chapter we stated the following result of Helton and Howe: If $T = X + iY$ is a nearly normal operator, then there is a regular signed measure P_T having compact support in $\mathbb{R}^2$ which provides the representation

$$\begin{aligned}(p,q) &= \operatorname{tr} i[p(X,Y),q(X,Y)] \\ &= \iint_{\mathbb{R}^2} J(p,q)\,dP_T ,\end{aligned} \tag{5.1}$$

where p,q are elements in $\mathcal{P}(\mathbb{R}^2)$. Recently, C. Berger [1] has obtaine estimates on the measure P_T that lead to most interesting results concerning invariant subspaces for seminormal operators. In this section, we will derive some of Berger's results.

Theorem 5.1. Let $T = X + iY$ and $A = U + iV$ be the Cartesian forms of two nearly normal operators T, A on the Hilbert space $\mathcal{H}$. Denote by P_T, P_A the Helton-Howe measures providing the representation (5.1) for the operators T, A, respectively. Suppose there is a trace class operator W which is one-to-one and has dense range that satisfies

$$AW = WT \quad . \tag{5.2}$$

Then for any Borel set $F \subset \mathbb{R}^2$

$$P_A(F) \le P_T(F) \quad . \tag{5.3}$$

Proof. For any real t, we will let J_t be the projection of $\mathcal{H} \oplus \mathcal{H}$ onto the graph of tW. Thus

$$J_t = \begin{bmatrix} Z_t & Z_t(tW)^* \\ (tW)Z_t & (tW)Z_t(tW)^* \end{bmatrix} ,$$

where $Z_t = (I+t^2W^*W)^{-1}$.

The operators $B_t = (T\oplus A)J_t \, (t\in\mathbb{R})$ are nearly normal operators on $\mathcal{H} \oplus \mathcal{H}$. Indeed,

$$J_t = \begin{bmatrix} I & 0 \\ 0 & 0 \end{bmatrix} + Q_t = J_0 + Q_t ,$$

where Q_t is a trace class operator. It follows that

$$[B_t^*, B_t] = \begin{bmatrix} [T^*,T] & 0 \\ 0 & 0 \end{bmatrix} + N_t ,$$

where N_t is trace class. This establishes the fact that $B_t \, (t \in \mathbb{R})$ is nearly normal.

Let $B_t = X_t + iY_t (t \in \mathbb{R})$ be the Cartesian form of the operator B_t. We will denote by $P_t (t \in \mathbb{R})$ the Helton-Howe measure providing the representation

$$\operatorname{tr} i[p(X_t,Y_t), q(X_t,Y_t)] = \iint_{\mathbb{R}^2} J(p,q)\,dP_t .$$

First it will be shown that $P_t = P_T$, for all $t \in \mathbb{R}$. To conclude this it suffices to establish

$$\operatorname{tr}[(B_t^*)^n, (B_t)^m] = \operatorname{tr}[(T^*)^n, T^m] , \tag{5.4}$$

for $m,n = 1,2,\ldots$. In fact, equation (5.4) is equivalent to the identity

$$mn \iint (x-iy)^{n-1}(x+iy)^{m-1} dP_t = mn \iint (x-iy)^{n-1}(x+iy)^{m-1} dP_T ,$$

This implies $P_t = P_T$.

We now turn to the proof of (5.4). Assume m,n are held fixed. The intertwining identity (5.2) implies the graph of tW is invariant under the operator $T \oplus A$. Consequently,

$$(B_t)^m = (T\oplus A)^m J_t, \quad (B_t^*)^n = J_t (T^* \oplus A^*)^n .$$

It follows that

$$[(B_t^*)^n, B_t^m] = J_t (T^*\oplus A^*)^n (T\oplus A)^m J_t - (T\oplus A)^m J_t J_t (T^*\oplus A^*)^n .$$

Substitute the identity $J_t = J_0 + Q_t$, where Q_t is trace class, in this last expression. A short computation implies (5.4) and this completes the proof that $P_t = P_T (t \in \mathbb{R})$.

Let r,s be real valued polynomials on $\mathbb{R}$. For $t \in \mathbb{R}$, we introduce the operators

$$\tilde{B}_t = r(s(X_t)Y_t s(X_t))X_t r(s(X_t)Y_t s(X_t)) + i\, s(X_t)Y_t s(X_t) .$$

The reader will recognize $\tilde{B}_t$ as the smooth version of a "double cut-down" of the operator B_t. We use the notation:

$$S_t = s(X_t), \quad R_t = r(s(X_t)Y_t s(X_t)) .$$

The self-commutator of $\tilde{B}_t$ has the form

$$[\tilde{B}_t^*, \tilde{B}_t] = R_t S_t [B_t^*, B_t] S_t R_t . \qquad (5.5)$$

rom the inequality

$$[B_t^*, B_t] \geq J_t[(T \oplus A)^*, T \oplus A] J_t$$

nd equation (5.5), we have

$$\mathrm{tr}[\tilde{B}_t^*, \tilde{B}_t] \geq \mathrm{tr}(R_t S_t J_t [(T \oplus A)^*, T \oplus A] J_t S_t R_t). \qquad (5.6)$$

For the left-hand side of the inequality (5.6), we use the dentity

$$\mathrm{tr}[\tilde{B}_t^*, \tilde{B}_t] = 2 \iint_{\mathbb{R}^2} r^2(s^2(x)y) s^2(x)\, dP_t$$

$$= 2 \iint_{\mathbb{R}^2} r^2(s^2(x)y) s^2(x)\, dP_T , \qquad (5.7)$$

here we have made use of the equality $P_T = P_t (t \in \mathbb{R})$.

It is clear that

$$\text{s} - \lim_{t\to\infty} J_t = \begin{bmatrix} 0 & 0 \\ 0 & I \end{bmatrix}$$

d, consequently,

$$\text{s} - \lim_{t\to\infty} S_t = \begin{bmatrix} 0 & 0 \\ 0 & s(U) \end{bmatrix}$$

$$\text{s} - \lim_{t\to\infty} R_t = \begin{bmatrix} 0 & 0 \\ 0 & r(s(U)Vs(U)) \end{bmatrix} .$$

conclude,

$$\lim_{t\to\infty} \mathrm{tr}(R_t S_t J_t [(T\oplus A)^*, T\oplus A] J_t S_t R_t)$$

$$= \mathrm{tr}(r(s(U)Vs(U))s(U)[A^*,A]s(U)r(s(U)Vs(U)))$$

$$= 2 \iint_{\mathbb{R}^2} r^2(s^2(x)y) s^2(x)\, dP_A . \qquad (5.8)$$

Taking into account (5.6)-(5.8), we obtain

$$\iint_{\mathbb{R}^2} r^2(s^2(x)y)s^2(x)dP_T \geq \iint_{\mathbb{R}^2} r^2(s^2(x)y)s^2(x)dP_A . \quad (5.9)$$

Let $F = [a,b) \times [c,d)$ be a rectangle in $\mathbb{R}^2$ and s_n^2 (respectively, r_n^2) a sequence of polynomials converging pointwise boundedly to the characteristic function of $[a,b)$ (respectively, $[c,d)$). Using these sequences $\{s_n\},\{r_n\}$ and taking the limit in (5.9), we obtain

$$P_A(F) \leq P_T(F) \quad .$$

The inequality (5.3) is now easily obtained for arbitrary Borel sets in $\mathbb{R}^2$. This completes the proof of the theorem.

The result in the above theorem remains valid if one assumes that W is Hilbert-Schmidt. The extra argument required for the Hilbert-Schmidt case is similar to the proof of Lemma 3.1 in Chapter 3.

The above theorem has the following corollaries:

Corollary 5.1. Let $T = X + iY$ be the Cartesian form of a nearly normal operator on $\mathcal{H}$ and let P_T be the Helton-Howe measure providing the representation (5.1). Suppose T and T^* have cyclic vectors. Then P_T is absolutely continuous with respect to planar Lebesgue measure. Moreover,

$$2\pi \left| \frac{dP_T}{dxdy} \right| \leq 1. \quad (5.10)$$

Proof. Without loss of generality, it can be assumed that $\|T\| < 1$. As we saw in the proof of Theorem 3.1 in Chapter 3, there is a trace class (see the remark following the proof of Theorem 3.1) operator $W: \ell_2^+ \to \mathcal{H}$ with dense range so that

$$TW = WU_+ \quad .$$

The operator W fails to be one to one if and only if $\varphi(T) = 0$ for some function φ analytic in the unit disc. The last statement is explained in the proof of Theorem 3.1 of Chapter 3. In this case, by a change of variables argument, we conclude the measure P_T is supported on the finite set $\{z:\varphi'(z) = 0, |z| \leq \|T\|\}$. However, as observed in Helton and Howe [1, p.189], this implies $P_T = 0$. As a result, we may assume W is one-to-one.

Directly from Theorem 5.1, we obtain

$$P_T(F) \leq P_{U_+}(F) = \frac{1}{2\pi} \iint_F g_{U_+} dxdy ,$$

where $F \subset \mathbb{R}^2$ is an arbitrary Borel set. From the fact that $[U_+^*, U_+]$ has rank one, we conclude $g_{U_+} \leq 1$. (See, e.g., Remark 2° in Section 3). It follows that

$$2\pi P_T(F) \leq \text{meas}_2(F) . \tag{5.11}$$

If we now apply the above argument to the operator T^*, we conclude

$$2\pi P_{T^*}(F) = -2\pi P_T(F) \leq \text{meas}_2(F) . \tag{5.12}$$

Inequality (5.10) follows from (5.11) and (5.12). This completes the proof.

The proof of the above corollary is interesting in that it does not presuppose the absolute continuity of the measure P_T. A more general result for the case of "multicyclic" operators appears in Berger's paper [1].

In order to derive the next result we need the following: Let H be a nearly normal hyponormal operator and g_H its principal function. It was remarked above (See, Remark 6° in Section 3) that the principal function g_{H^n} $(n = 1,2,\ldots)$ is given by the identity

$$g_{H^n}(x,y) = \sum_{(\xi+i\eta)^n = x+iy} g_H(\xi,\eta) .$$

It is easy to conclude when H is non-normal that for n sufficiently large

$$\operatorname{esssup}(g_{H^n}) > 1.$$

This leads to the following result of Berger [1].

Corollary 5.2. Let H be a non-normal, seminormal operator. Then for some integer N the operator $H^n (n \geq N)$ fails to have a cyclic vector.

The above corollary follows directly from Theorem 5.1. It should be noted that Corollary 5.2 is false for cohyponormal operators. In fact, the operators $(U_+^*)^n (n \geq 1)$ all have many cyclic vectors (See, Halmos [1]). It is possible to formulate a result along the lines of Corollary 5.2 for a nearly normal operator T. In fact, if $P_T \neq 0$, then there is $z \in \mathbb{C}$ and $n > 0$ such that either T_z^n or T_z^{*n} fails to have a cyclic vector.

We can state another corollary of Theorem 5.1:

Corollary 5.3. Let $E \subset \mathbb{R}$ be a bounded measurable set. For $n \geq 2$, let B be a non-negative element of $L^\infty(E:M_n)$. The hyponormal singular integral operator S defined on $L^2(E:\mathbb{C}^n)$ by

$$Sf(x) = xf(x) + \frac{B(x)}{\pi} \int_E \frac{B(t)f(t)}{x-t}\,dt$$

has a non-trivial invariant subspace. Further, if

$$\det B(x) \neq 0 \tag{5.13}$$

for all x in some set of positive measure, then the operator S fails to have a cyclic vector.

Proof. The matrix function B can be diagonalized measurably. This means there is a $U \in L^\infty(E:M_n)$, which is unitary valued a.e., such that

$$B = UDU^* ,$$

vhere $D(x) = \mathrm{diag}[b_1(x), \ldots, b_n(x)]$. The numbers $b_1(x), \ldots, b_n(x)$ are the eigenvalues of $B(x)$.

If (5.13) does not hold, then for some j, $b_j(x) \equiv 0$ on a set $' \subset E$ of positive measure. The operator S is unitarily equivalent to the operator

$$S'f(x) = xf(x) + \frac{D(x)U^*(x)}{\pi} \int_E \frac{U(t)D(t)f(t)}{x-t}\, dt \quad .$$

et $\mathfrak{m}$ be the subspace of $L^2(E:\mathbb{C}^n)$ consisting of functions $(f_1,\ldots,f_n)$ uch that f_j vanishes on F. The subspace $\mathfrak{m}$ is clearly a reducing subpace for S'. Consequently, in case (5.13) fails to hold, the perator S has a reducing subspace.

If (5.13) holds, then the principal function g_S of the operator S ill be at least two on some set of positive measure. This follows rom the explicit description of g_S given in Proposition 5.2. The esult in Theorem 5.1 implies the operator S does not have a cyclic ector. The proof is complete.

Notes

Section 1. The study of almost commuting pairs of operators via ne tracial bilinear form originates in the work of Helton and Howe [1]. ne principal function approach to an almost commuting pair began ith the work of Pincus [1] concerning the diagonalization of selfljoint singular integral operators. The principal function has been udied by Carey and Pincus [6] for a pair of self-adjoint elements in von Neumann algebra equipped with a normal semi-finite trace. The alogue of Theorem 1.1 for tuples of almost commuting operators is udied in Helton and Howe [2].

Section 2. The proof of the absolute continuity of the tracial linear form is in Carey and Pincus [3]. There is another interesting principal function that has been associated with a nearly normal operator. This is the principal function of the pair consisting of the factors in the polar factorization of the operator. The polar principal function is discussed in Carey and Pincus [1,6]. In the case where the self-commutator of the nearly normal operator has rank one, the principal function becomes a complete unitary invariant. In this situation, Carey and Pincus [7] have constructed an interesting canonical model based on the principal function.

Section 3. Most of the properties of the principal function discussed in this section are in Carey and Pincus [6] and Helton and Howe [1]. In each of these works the authors develop a fairly flexible functional calculus. Thus, for example, the change of variables formula can be extended beyond the case of polynomials.

Section 4. The examples in this section go back to the work of Pincus [1].

Section 5. The work of Berger [1], which is briefly discussed in this section, should serve as a springboard for much new work concerning nearly normal operators.

Bibliography

Aronszajn, N. and W. F. Donoghue, Jr.:

[1] On exponential representations of analytic functions in the upper half-plane with positive imaginary part, Jour. d'Analyse Math. 5 (1956-57), 321-388.

Apostol, C. and K. Clancey :

[1] Local resolvents of operators with one-dimensional self-commutator, Proc. Amer. Math. Soc. 58 (1976), 158-162.

Berger, C. A.:

[1] Sufficiently high powers of hyponormal operators have rationally invariant subspaces, Journal of Integral Equations and Operator Theory 1 (1978), 444-447.

Berger, C. A. and B. I. Shaw:

[1] Hyponormality its analytic consequences, Amer. J. of Math (to appear).

Bram, J.:

[1] Subnormal operators, Duke Math. J. 22 (1955), 75-94.

Brown, L.:

[1] The determinant invariant for operators with trace class self-commutator, Proceedings of a conference on operator theory, Springer Verlag Lecture Notes no. 345, 1973.

Brown, S.:

[1] Some invariant subspaces for subnormal operators, Journal of Integral Equations and Operator Theory 1 (1978), 310-333.

Carey, R. W.:

[1] A unitary invariant for pairs of self-adjoint operators, J. Reine Angew. Math. 283 (1976), 294-312.

Carey, R. W. and J. D. Pincus:

[1] The structure of intertwining isometries, Indiana Univ. Math. J. 22 (1973), 679-703.

[2] An invariant for certain operator algebras, Proc. Nat. Acad. Sci. USA 71 (1974), 1952-1956.

[3] An exponential formula for determining functions, Indiana Univ. Math. J. 23 (1974), 1031-1042.

[4] Construction of seminormal operators with prescribed mosaic, Indiana Univ. Math. J. 23 (1974), 1155-1165.

[5] Commutators, symbols and determining functions, J. Functional Analysis 19 (1975), 50-80.

[6] Mosaics, principal functions, and mean motion in von Neumann algebras, Acta. Math. 138 (1977), 153-218.

[7] The structure of intertwining partial isometries, II; Canonical models, preprint.

Clancey, K.:

[1] Seminormal operators with compact self-commutators, Proc. Amer. Math. Soc. 26 (1970), 447-454.

[2] On the local spectra of seminormal operators, Proc. Amer. Math. Soc. 72 (1978), 473-479.

Clancey, K. and C. R. Putnam:

[1] The spectra of hyponormal integral operators, Commentarii Math. Helvetici 46 (1971), 451-456.

Conway, J. B. and R. F. Olin,:

[1] A functional calculus for subnormal operators II, Memoirs of the A.M.S., Vol. 184 (1977).

Diximier, J.:

[1] Les Algebres d'Operateurs dans l'Espace Hilbertien, Gauthier-Villars, Paris, 1969.

Donoghue, W. F. Jr.:

[1] On the perturbation of spectra, Comm. on Pure and Applied Math. Vol. XVIII (1965), 559-580.

Douglas, R. G.:

[1] On majorization, factorization and range inclusion of operators on Hilbert space, Proc. Amer. Math. Soc. 17 (1966), 413-415.

Duren, P. L.:

[1] Theory of H^p Spaces, Academic Press, New York, 1970.

Friedrichs, K. O.:

[1] On the perturbation of continuous spectra, Comm. App. Math. 1 (1948), 361-406.

arsia, A.:

[1] Topics in Almost Everywhere Convergence, Markham, Chicago, 1970.

ohberg, I. C. and M. G. Krein:

[2] Introduction to the Theory of Linear Nonself-adjoint Operators, Amer. Math. Soc., Providence, Rhode Island, 1969.

almos, P. R.:

[1] A Hilbert Space Problem Book, Van Nostrand, Princeton, N. J., 1967.

elton, J. W. and R. Howe:

[1] Integral operators, commutator traces, index and homology, Proceedings of a conference on operator theory, Springer Verlag Lecture Notes No. 345, 1973.

[2] Traces of commutators of integral operators, Acta Math. 135 (1975), 271-305.

errero, D. A.:

[1] Interpolation between two Putnam's inequalities, Rivista de la Union Mat. Argentina 28 (1976), 42-45.

owe, R.:

[1] A functional calculus for hyponormal operators, Indiana Univ. Math. J. 23 (1974), 631-644.

ohnson, B. E.:

[1] Continuity of linear operators commuting with continuous linear operators, Trans. Amer. Math. Soc. 128 (1967), 88-102.

ato, T.:

[1] Perturbation Theory for Linear Operators, Springer Verlag, N. Y., 1966.

[2] Smooth operators and commutators, Studia Math. 31 (1968), 535-546.

rein, M. G.:

[1] Perturbation determinants and a formula for the traces of unitary and self-adjoint operators, Dokl. Akad. Nauk. SSSR 144 (1962), 268-271.

roda, S. T.:

[1] On a generalization of the Weinstein-Aronszajn formula and the infinite determinant, Sci. Papers Colloq. Gen. Education, Univ. of Tokyo 11 (1961), 1-12.

Loewner, K.:

[1] Uber monotone matrix funktionen, Math. Zeits. 38 (1934), 177-216.

Muhly, P. S.:

[1] A note on commutators and singular integrals, Proc. Amer. Math. Soc. 54 (1976), 117-121.

Pincus, J. D.:

[1] Commutators and systems of singular integral equations, I, Acta Math. 121 (1968), 219-249.

[2] The spectrum of seminormal operators, Proc. Nat. Acad. Sci. USA 68 (1971), 1684-1685.

[3] On the trace of commutators in the algebra of operators generated by an operator with trace class self-commutator, Stony Brook preprint (1972).

[4] The determining function method in the treatment of commutator systems, Colloquia Math. Soc. Janos Bolyai 5, Hilbert space operators, Tihany (Hungary) (1970), 443-477.

Putnam, C. R.:

[1] Commutation Properties of Hilbert Space Operators and Related Topics, Ergebnisse Der Math. 36, Springer, New York, 1967.

[2] Commutators, absolutely continuous spectra, and singular integral operators, Amer. Jour. Math. 86 (1964), 310-316.

[3] On the spectra of seminormal operators, Trans. Amer. Math. Soc. 119 (1965), 509-523.

[4] The spectra of seminormal singular integral operators, Can. J. Math. XXII (1970), 134-150.

[5] An inequality for the area of hyponormal spectra, Math. Z. 28 (1971), 473-477.

[6] Ranges of normal and subnormal operators, Mich. Math. Jour. 18 (1971), 33-36.

[7] A similarity between hyponormal and normal spectra, Ill. Jour Math. 16 (1972), 695-702.

[8] Resolvent vectors, invariant subspaces and sets of zero capacity, Math. Ann. 205 (1973), 165-171.

[9] The role of zero sets in the spectra of hyponormal operators Proc. Amer. Math. Soc. 43 (1974), 137-140.

[10] Hyponormal contractions and strong power convergence, Pac. Jour. of Math. 57 (1975), 531-538.

[11] Hyponormal operators and spectral multiplicity, preprint.

[12] Invariant subspaces of operators having nearly disconnected spectra, preprint.

Radjabalipour, M.:

[1] Ranges of hyponormal operators, Ill. J. Math. 21 (1977), 70-75.

[2] On majorization and normality of operators, Proc. Amer. Math. Soc. 62 (1977), 105-110.

Riesz, F. and Sz-Nagy, B.:

[1] Functional Analysis, Fredrick Ungar Pub. Co., New York (1955).

Rosenblum, M.:

[1] A spectral theory for self-adjoint singular integral operators, Amer. Jour. of Math. 88 (1966), 314-328.

Rudin, W.:

[1] Real and Complex Analysis, Second Edition, McGraw-Hill, New York, 1974.

Saks, S.:

[1] Theory of the Integral. (Second revised edition) English translation by L. C. Young. G. E. Stechert and Co., New York, 1937.

Schatten, R.:

[1] Norm Ideals of Completely Continuous Operators, Springer, Berlin (1960).

Schwartz, J.:

[1] W^*-algebras, Gordon and Breach, New York (1967).

Stampfli, J. G.:

[1] Hyponormal operators, Pac. Jour. Math. 12 (1962), 1453-1458.

[2] Hyponormal operators and spectral density, Trans. Amer. Math. Soc. 117 (1965), 469-476.

[3] Analytic extensions and spectral localization, Jour. of Math. and Mechanics 16 (1966), 287-296.

[4] A local spectral theory for operators, J. Functional Analysis 4 (1969), 1-10.

[5] A local spectral theory for operators II, Bull. Amer. Math. Soc. 75 (1969), 803-806.

[6] A local spectral theory for operators III: Resolvents, spectral sets and similarity, Trans. Amer. Math. Soc. 168 (1972), 133-151.

[7] A local spectral theory for operators IV: Invariant subspaces, Indiana U. Math. J. 22 (1972), 159-167.

[8] A local spectral theory for operators V: Spectral subspaces for hyponormal operators, Trans. Amer. Math. Soc. 217 (1976), 285-296.

Stampfli, J. G. and B. L. Wadhwa:

[1] On dominant operators, Monatsh. Math. 84 (1978), 143-153.

Titchmarsh, E. C.:

[1] Introduction to the Theory of Fourier Integrals, Clarendon Press, Oxford, 1937.

Xa Dao-xeng (Hsia Tao-hsing):

[1] On non-normal operators, Chinese Math. 3 (1963), 232-246.

INDEX